The Earth
Moves

GREAT DISCOVERIES

DAN HOFSTADTER

The Earth Moves

Galileo and
the Roman Inquisition

ATLAS & CO.

W. W. NORTON & COMPANY

NEW YORK · LONDON

For information about permission to reproduce selections from
this book, write to Permissions, W. W. Norton & Company, Inc.,
500 Fifth Avenue, New York, NY 10110

For information about special discounts for bulk purchases, please contact
W. W. Norton Special Sales at specialsales@wwnorton.com or 800-233-4830

Manufacturing by RR Donnelley, Bloomsburg
Book design by Chris Welch
Production manager: Julia Druskin

Library of Congress Cataloging-in-Publication Data

Hofstadter, Dan.
The Earth moves : Galileo and the Roman Inquisition /
Dan Hofstadter. — 1st ed.
p. cm. — (Great discoveries)
Includes bibliographical references and index.
ISBN 978-0-393-06650-0 (hardcover)
1. Galilei, Galileo, 1564-1642—Trials, litigation, etc. 2. Inquisition—Italy—Rome.
3. Astronomy—Religious aspects—Christianity—History of doctrines—17th
century. 4. Science, Renaissance. 5. Catholic Church—Doctrines—History—17th
century. 6. Catholic Church—Italy—History—17th century. I. Title.
QB36.G2H64 2009
509.4'09032—dc22

2009004325

ISBN 978-0-393-33820-1 pbk.

W. W. Norton & Company, Inc.
500 Fifth Avenue, New York, N.Y. 10110
www.wwnorton.com

W. W. Norton & Company Ltd.
Castle House, 75/76 Wells Street, London W1T 3QT

1 2 3 4 5 6 7 8 9 0

For Bette

Contents

The Earth
Moves

There are many books about Galileo, so the reader is entitled to ask how this one differs from any others that he or she might come across. It is, as the title suggests, an attempt to recount how the great physicist and astronomer Galileo Galilei was tried and convicted by the Roman Inquisition in 1633 for championing the Copernican hypothesis about the solar system, against which the Vatican had issued an edict. I have tried to tell this story succinctly for the general reader, and without inserting any but a few paragraphs of the simplest mathematics into the text.

In addition, I have chosen a very specific path through a thicket of information. Galileo had several reasons for embracing Copernicanism. Deductions in both mechanics and mathematical astronomy contributed to his growing conviction that the earth revolved around the sun, and I have touched upon these; I have devoted much more attention, however, to his sudden and vast improvement of the telescope in the autumn of 1609, which enabled him to actually see the

surface of the moon, the phases of Venus, and the moons of
Jupiter, and so to behold celestial mechanics at work. In time
these telescopic sightings would put him at greater variance
with the edicts of the Roman Catholic Church than his far
greater discoveries in physics ever did. This was the first great
clash of religion and science, and it still has much to teach us.

What Galileo did with the telescope is in itself an exciting
story. But I have a special reason for retelling it. I am primarily
interested in the arts, and Galileo loved music, literature, and
painting. A musician's son, he played the lute well, wrote
poetry and literary criticism, taught perspective, and drew
with some verve; he corresponded at length with at least one
major painter. He favored the classics, yet he belonged to a
world no longer classical but Baroque in orientation. A system
of thought may be described as Baroque when its parts cannot
be understood or enjoyed unless they are constantly related to
some larger, dynamic whole, and in that sense Galileo was a
prime representative of the Baroque era, which began around
1600. Geometry, trigonometry, and perspective were then still
the common property of mathematicians, painters, and architects, and the budding science of optics interested them all:
indeed we find painters among the most enthusiastic supporters of Galileo's discoveries with the telescope. Understandably,
Galileo's position within the general context of Baroque civilization has not much concerned historians of science, and I
have tried to offer a brief picture of it here.

Most writing about Galileo and the Inquisition has to do
with philosophy, and that is as it should be. But there are
already a number of excellent books on this subject, and I
found myself more drawn to the psychological complexities of
the 1633 trial. I have always found it hard to control my bewil-

derment over the fact that Galileo's persecutor, Pope Urban VIII, is known in art history as Bernini's and Borromini's great patron, indeed as the most devoted supporter that the Baroque style ever found in Italy; during the 1620's he also befriended and encouraged Galileo. In this book I have tried to accord him more sympathy than he usually receives.

The man who really caught my attention, however, was the all-but-forgotten Tuscan ambassador to the Vatican, Francesco Niccolini, who is the virtual hero of many of the pages that follow. Since Galileo was mathematician to the court of the Grand Duke of Tuscany, both the duke and his emissary Niccolini wished to avert a trial by the Inquisition, or, failing that, to curtail an eventual trial by means of an extra-judicial solution—what we would call a plea-bargain. Unhappily Galileo, as one of Europe's first professional intellectuals, did not fit into this scheme. Unlike earlier Italians who had taught largely by personal example—unlike, say, Saint Francis, who kissed the leper, or Giordano Bruno, who chose to die rather than think as he was told to—Galileo had faith in the transcendent value of a good argument, and he wanted to argue with the Vatican. He even wanted to argue with the Inquisition. Ambassador Niccolini, with his insight into the papal court, foresaw the danger of such a project and tried to induce Galileo to abandon it. Reading Niccolini's correspondence, I found myself touched by the abrasion between two friends representing such different ethical approaches to life.

Galileo has often been described as an inordinately suspicious man. There is some truth to this, but his friends' letters suggest that he was also subjected to a great deal of envy and hostility. The literary background to the concept of envy, or *invidia*, was raised some years ago by Miles Chappell in relation to two graphic works by Ludovico Cigoli, one of them dedicated to his friend Galileo. I have returned to this idea, stressing its background in Dante's *Divine Comedy*.

While on the subject of my debt to Galilean scholars, I wish to express how much I owe to my reading of Stillman Drake and William R. Shea (who has particularly influenced me). Much of the material on the telescope has benefited from my reading of Albert Van Helden's papers on the subject. I confess to an almost hypnotic fascination with the philosophical dimension of the trial, and here I must gratefully mention Richard J. Blackwell, Annibale Fantoli, Rivka Feldhay, Maurice A. Finocchiaro, Ernan McMullin, and Guido Morpurgo-Tagliabue, among others. Certain views expressed in this book accept Eileen Reeves's conclusions about Ludovico Cigoli in her *Painting the Heavens* (although I think that morphologically Galileo's thought is much more closely related to Bernini's). Precise references to all these writers' works can be found in the bibliography at the end of this book.

The literature on Galileo is vast, and I have surely overlooked many books and essays. I should like to make particular mention of two works that came to my attention when I had already turned in this manuscript to the editors. One is Horst Bredekamp's *Galileo Der Künstler: Der Mond. Die Sonne. Die Hand*, which contains some fascinating revelations concerning Galileo's manual involvement in fashioning pictures of the heavens. The other is Dr. Giorgio Strano's *Il Telescopio di Galileo: lo strumento che ha cambiato il mondo*, which I have so far been unable to procure.

Where I have cited translations of various texts, the translators are credited in the notes. The translation from Ariosto's *Orlando furioso* is by David Slavitt. The translations from

Dante, and the English version of the letters in colloquial Italian relative to the 1633 trial, are by myself.

FOR HELP ALONG the way, my heartfelt thanks go to Barbara Dudley, for enabling me to achieve a secure internet connection at a time when this was difficult in our rather secluded village; to David Slavitt, for his delightful, as-yet-unpublished Ariosto translation; to Nino Mendolia, for his work on the illustrations; to Prof. Norman Derby, for his clarification of several problems in astronomy; to Jim Mosher and Tim Pope, for their kind permission to use their excellent ray diagrams, and to Jim Mosher in particular for his explanations of various points relative to Galilean optics; to Dr. Giorgio Strano, of the Istituto e Museo della Storia delle Scienze, in Florence, for consenting to be interviewed at the institute about his research into Galileo's telescope during the summer of 2007, before the publication of his research; to Prof. Ricardo Nirenberg, that extraordinary polymath, for our many conversations, and for carefully reading the manuscript and notifying me about a number of errors; to Prof. Glen Van Brummelen, also for reading the manuscript from the standpoint of the history of mathematics; to Prof. William R. Shea, of the University of Padua; and to Oceana Wilson, of the Bennington College Library, Alessandra Lenzi, of the Biblioteca dell'Istituto e del Museo della Storia delle Scienze, and Mary DiAngelo, of the Schow Science Library of Williams College, for their assistance in procuring me many books and papers.

The Summons

Galileo Galilei was living in a modest house in Arcetri, in the hills south of Florence, when he learned, on October 1, 1632, that he had been summoned to Rome to be examined by the Inquisition. Arcetri now lies in the heart of what you might call *la Firenze bene*—upper-class suburban Florence. By chance I spent two years in the same area and would often walk my baby son through the village in his stroller, thinking sometimes about Galileo, who wrote a great work about physics here, or about the poet Eugenio Montale, whose ashes are preserved in a cemetery nearby. The southern Florentine countryside is a soothing spectacle, and never more so than at this season—harvest time. Every hilltop affords a vista of rolling horizon, with an occasional disclosure of the dome of the Cathedral of Santa Maria del Fiore, and from among the tender gray-green of the olive orchards and the richer green of the vineyards rise the bell towers of ancient monastic establishments, such as the Charterhouse of Galluzzo and the Convent of the Stigmatine Sisters (whose

playground my little son particularly favored). In another
Franciscan convent, San Matteo, only minutes away by foot,
Galileo's elder daughter, Suor Maria Celeste, then thirty-two,
served as a nun. It is hard to stroll through this landscape
remembering the Inquisition's summons without being struck
by the contrast between the tranquil loveliness of the place
and the extreme anguish that the news provoked in Galileo,
his family, and his friends.

The immediate cause of the summons was the failure of the
grand duke of Tuscany to get the matter dismissed or at least
transferred to the office of the Roman Inquisition in Florence.
But the underlying cause was a book that Galileo had published in May of 1632, the *Dialogue Concerning the Two Chief
World Systems*, in which subtle arguments in favor of the geocentric, or earth-centered, and the heliocentric, or sun-
centered, cosmologies were played against each other.
Organized somewhat like a Platonic dialogue, it featured lucid
explanations, crisp Italian prose, and entertaining arguments
between three well-defined characters. The Roman Catholic
Church at this time had no dogma regarding the structure of
the universe, but of course it had a theology, and the Council
of Trent, which ended in 1563, had confirmed that the Bible,
which stated that the sun rose in the east and set in the west,
was not to be freely interpreted by laymen. Galileo's book had
been approved by the Vatican censor, but a number of clerics,
especially among the Jesuits, still felt that it inclined too heavily toward Copernicus, the great heliocentric astronomer. In
the summer of 1632, a commission was empanelled to study
the book, and by late September the work was provisionally
suppressed. Pope Urban VIII Barberini was an old friend and
admirer of Galileo's, but he was nonetheless known to be

incensed by the Tuscan scientist's alleged infringement of restrictions as to what might be legitimately argued, as well as by other audacious features of the *Dialogue*. It was beneath the pope to summon Galileo himself, and he apparently had trouble finding other churchmen willing to do so. In the end, it fell to his brother Antonio, a morbidly timid Capuchin monk whom Urban had all but forced to become a cardinal, to notify Galileo that his presence was expected in Rome.

Galileo was then sixty-nine years old. The turning point in his life had come in late 1609, when he was forty-five, a respected mathematician with a sideline in engineering. That autumn, he had rapidly improved upon a crude spyglass circulating throughout Europe, and within months, from atop his house in Padua, had seen the rugged surface of the moon, the phases of Venus, Jupiter's satellites, and myriads of shimmering stars previously unknown to humanity. For years he had mulled over his research in mechanics and astronomy and leaned toward the Copernican theory of the heavens. He had questioned the validity of the two competing hypotheses, the Ptolemaic, or geocentric, and the Tychonic, or geoheliocentric (which seemed so preposterous, so illogical, that he never took it very seriously). But after his telescopic observations he made up his mind—there was no doubt that the sun stood at the center of the cosmos. He did not publicly proclaim this fact, however. Only here and there, in his private correspondence or in certain published works, such as his *Letters on Sunspots* of 1613 or *The Assayer* of 1623, did he frankly state that Copernicus was right about the sun and his opponents wrong.

In 1616, the Congregation of the Holy Office, otherwise known as the Roman Inquisition, had banned any advocacy or teaching of the Copernican position, on the grounds that it conflicted with Scripture. Copernicus's exposition of his system, *On the Revolutions of the Heavenly Spheres*, of 1543, was to be "corrected" by the Vatican Index, that is, emended to indicate its purely hypothetical status. This notion of heliocentrism as a sort of counterfactual, mathematical fiction was evolving dramatically then, and as time went on Galileo interpreted it according to his own lights. Following the publication of *The Assayer*, and encouraged by the election of a new, friendly, humanist pope—Urban VIII Barberini—Galileo decided to write a magnum opus weighing the mutually opposing arguments for the geocentric and heliocentric theories. He was giddily spurred on in this task by a former student named Giovanni Ciampoli, now the pope's correspondence secretary and an influential backroom schemer, who had surprisingly little sense of the danger of the undertaking. Galileo resolved to keep his *Dialogue* on the plane of mere supposition, but, considering his deep scientific engagement on behalf of Copernicanism, this was scarcely possible.

In an orderly, logical world, Galileo, on discovering the truth about the heavens in 1609–12, would have published a tract defending Copernicanism, and this tract would have either won the Church over or provoked a scientific and theological rebuttal from the Collegio Romano, the great Jesuit scholarly institution in Rome. But as it turned out, the Collegio Romano enthusiastically confirmed almost all of Galileo's telescopic discoveries. In 1616, Cardinal Robert Bellarmine, the most gifted Catholic theologian and polemicist of the age and the director of the Roman Inquisition, merely warned Galileo not to go drawing any perilous conclusions from his observations. Bellarmine was a very polite man, and the admonition may have been rather too gentlemanly to produce the desired effect. And though Galileo was investigated by the Inquisition at that time, he was never deposed or indicted.* Thus it was not until the publication of the *Dialogue*, in 1632, that the scientist fell fatally afoul of Rome.

If one has not read the *Dialogue*, one might suppose it consists of a grand summary of everything that Galileo had *seen* and *proved* about the heavens, mostly thanks to his improved telescope. That is not the case. The *Dialogue* is a fascinating work, but less an empirical than a theoretical, and in fact speculative, defense of Copernicanism. It is based largely on physics and celestial mechanics, and one of its most important arguments—Galileo's belief that the sun's gravitational field causes the tides—is wrong. There are other, valid explanations in the book, relating to the mathematical description of falling bodies, the confutation of popular anti-Copernican arguments, and why a ball tossed in the air as the earth rotates does not descend some distance to the west. To anyone ignorant of such matters—and almost everyone was in 1632—this material makes lucid and thrilling reading. But anyone looking for a step-by-step statement of the logical connection between what Galileo had seen through the telescope and the setup of the solar system would have been disappointed.

* It is sometimes said that Galileo faced two trials, the first in 1615–16 and the second in 1633. This is true in Italian but not English, because the word *processo* is not exactly cognate with the English "trial." A *processo* may begin, not from the moment a tribunal is convened, but as soon as a prosecutorial investigation of a suspect is initiated.

What one must bear in mind, then, is that in 1632–33 the Inquisition was not out to disprove Galileo's science or even to pinpoint all the passages of the *Dialogue* that conflicted with Scripture. The question was actually quite different. Christians had been warned not to teach Copernicanism: had Galileo heeded that warning or defied it? All the rest was beside the point.

There is thus an enormous distinction between the proximate cause of the Galileo affair, the great religion-science clash of 1633 that in some form has persisted into our time, and what really brought it about. The proximate cause was the publication of a work that appeared to flout a prohibition against Copernicanism. What brought it about was the evolution of a new science, an entire attitude toward experience that emphasized the evidence of the senses, especially vision, as opposed to metaphysics and the study of venerated texts. Over and over in his writings, Galileo expresses an almost petulant amazement that people will not *see* what nature so beautifully displays, nor search for the transcendent forms governing that beauty. In *The Assayer* this plea takes poetic flight. Perhaps because his father was a musician, and as a child he was trained as a lutenist, Galileo tells a parable about a lover of birds and birdsong who desires to look into all the ways that musical pitches are produced, whether by birds, people, or insects. This man travels the world collecting flutes, violins, even squeaky hinges and crystal goblets that produce a tone when filled with water. He examines wasps, mosquitoes, trumpets, organs, fifes, jews' harps, until at length he captures a cicada. Galileo tells us that the man, scrutinizing the cicada, failed to diminish its strident noise either by closing its mouth or stopping its wings, yet he could not see it move the scales that covered its body, or any other thing. At last he lifted up the armor of its chest and there he saw some thin hard ligaments beneath; thinking the sound might come from their vibration, he decided to break them in order to silence it. But nothing happened until his needle drove too deep, and transfixing the creature he took away its life with its voice, so that he was still unable to determine whether the song had originated in those ligaments. . . . I could illustrate with many more examples Nature's bounty in producing her effects, as she employs means we could never think of without our senses and our experiences to teach them to us—and sometimes even these are insufficient to remedy our lack of understanding. . . . The difficulty of comprehending how the cicada forms its song even when we have it right in our hands ought to be more than enough to excuse us for not knowing how comets are formed at . . . immense distances.

If there is anything that Galileo championed it was curiosity—immense, boundless curiosity.

There are many valid ways of presenting Galileo's thought. In a short book like this, it seems to me most useful to offer, as a background to the trial of 1633, a concise account of how he brought optics into alignment with astronomy in order to see and know the heavens. Because his telescopic observations made him a convert to Copernicanism, they formed the substantive reason for his divergence from the thought patterns prescribed by the Church, and it was this divergence, between

1609 and 1632, that caused the clash between religion and science. I also devote more attention to Galileo's relations with painters and his pictorial representation of astronomical facts, and to the Church's use of the frescoed dome as an alternative celestial vision, than might ordinarily be expected. In part this reflects a personal interest; in part it embodies my belief that the sense of vision, of *seeing* as opposed to the *refusal to see*, was enormously important to Galileo. I mean this more in the figurative than the literal sense, yet Galileo is reported to have drawn well, and he was friendly with a number of artists, especially Ludovico Cigoli, who ardently supported him and with whom he corresponded regularly. Most of those who have written about his relations with artists, such as Erwin Panofsky and Eileen Reeves, have dwelt on this connection with Cigoli, which was certainly paramount from a biographical standpoint. I do not think it should obscure the fact that Galileo's conception of space has much more in common with that of Bernini, whom I do not believe he ever met.

Galileo had a good grasp of Latin and some Greek and reportedly knew by heart many long passages penned by Virgil, Ovid, Dante, Petrarch, and Ariosto. He wrote poetry, a play, voluminous letters, and numerous essays and books on physics and astronomy. When he was upset, his syntax became turgid. Otherwise he wrote a straightforward, cogent Italian, by turns playfully humorous and bitingly ironic. His prose is so free of Latinisms and ornaments, so similar to the best of modern science writing, that it has been included in textbook surveys of the history of prose composition for Italian high schools. Galileo sometimes tackles with ordinary language scientific problems that contemporary physicists would approach with algebraic symbols: he is one of the last scientists who also seems something of an artist (Milton called him "the Tuscan artist"). As such, he makes frequent use of metaphor, both consciously and unconsciously. One suspects that this use of metaphor got him into trouble.

Of course, science doesn't need metaphor, and though metaphors may be embedded in scientific terms, such as "wheel" in *kúklos* (Greek for "circle"), or "overshooting" in *huperbolé*, or "comparison" in *parabolé*, geometry would likely have been the same had they had other names. But often, as in the case of Galileo, Darwin, and Freud, scientists or their supporters enlist a set of durable metaphors that, however dispensable for science itself, somehow fascinate the public and inflame its religious segment. Such has been the case with "plurality of worlds," "laws of nature," "descent of man," "survival of the fittest," and "Oedipus complex," heuristic terms that have caught on, perhaps lamentably, as catchwords. Often it is nascent sciences, or those in periods of theory shift, that rely for a while on metaphor. (Nowadays researchers in cognitive psychology may speak of information being "encoded" or "indexed" in a subject's memory.) Sometimes Galileo deliberately enlists metaphors, such as "the book of nature," or the "repugnance" of a body to being pulled in a certain direction. At other times he violates unspoken metaphorical conventions, and this can happen almost by accident. When one denies, as he did, that the moon is a crystalline sphere, and asserts that it is a spiky, pitted ball, one appears to be making a factual statement, but that is not all one is doing: one is also offending the Marian poetic vision that associates the Madonna with the unsullied moon. I have tried to make the metaphorical abrasion between Galileo and his clerical opponents clearer than it might appear in a strictly scientific context.

All the books on Galileo's trial that I have read present his chief antagonist, the pope, in his role as prelate, theologian, political leader, and string-puller of the Inquisition. The portrait is never flattering, even when penned by conservative Catholics. But the annals of art present a quite different picture of Urban VIII Barberini, and I have tried to merge it with the one offered by the history of science. It will come as no surprise to anyone with an interest in the Baroque style that Maffeo Barberini was a great patron of Italian sculpture and architecture. A highly cultivated man, he was also a keen follower of developments in science, and that he should have been the one to silence Galileo in the trial of 1633 was a calamity for him as well as for the Church.

1

Galileo Galilei and
Maffeo Barberini

alileo Galilei was born in Pisa in 1564, to the family of
a professional musician who later found employment
in Florence. According to his first biographer, Vincenzo Viviani, he was fond of drawing and originally wished to
become a painter. His father directed him toward a career in
medicine, but after a few months at the University of Pisa the
young man abandoned his medical studies and enrolled in the
faculty of mathematics. This field then included a hodgepodge
of subjects related to number and magnitude, such as optics
and fortification, but for Galileo, mathematics meant Euclid.
The preference was prophetic. Euclidean geometry, more than
any other subject, teaches one how to think—to think logically,
deductively: it is the essence, the bread of thought. At the same
time, in the absence of algebra, with which Galileo was unac-
quainted, Euclidean geometry provided the readiest tool with
which to grasp and conceive of spacial relationships. Galileo
soon proved himself an agile geometer, and was granted the
mathematics chair first at Pisa, from 1589 to 1592, and then at

Padua, from 1592 to 1610. And it was geometry that he began
to apply, as dexterously as anyone could, to such subjects as the
velocity of falling and rolling bodies.

While teaching at Pisa, Galileo began to study Archimedes,
many of whose surviving writings had been republished in
Basel in 1544. What he learned from the ancient Syracusan was
a way of conceptualizing the world in terms of intellectual
machines or models, such as the lever and the balance. This
way of thinking accorded well with his love of geometry and
his essentially Platonic cast of mind. He saw the world in terms
of Euclidean forms that not only held universally true but also
were quantifiable, in the sense that extensions or areas may be
proved greater or lesser than others. In Viviani's biography one
reads of an experiment in which Galileo drops unequal
weights off the Tower of Pisa in order to prove the equality of
their velocity of fall, but the story is almost certainly untrue:
while at Pisa he wrote (but did not publish) a manuscript on
physics, titled *De motu* (*On Motion*), in which he argued something altogether different about falling bodies, which suggests
that he had not made the experiment. Whether during this
period he even conducted experiments, and how numerous
they may have been, is still a matter of scholarly conjecture.
Early on, he worked on what became his law of fall, which
states that in a vacuum all bodies are uniformly accelerated,
and that the distance fallen is proportional to the square of the
elapsed time; but measuring fall was impossible then, as—
leaving aside the impossibility of creating a vacuum—bodies
fall too fast, and no means of retarding them by counter-
weights had yet been contrived. Galileo devised geometrical
thought-experiments extrapolated from actual experiments,
which he carried out with metal balls on inclined planes.

It is hard for us today to imagine two oppressive conditions under which the young Galileo worked. One was the dead hand of Aristotelian dynamics, universally taught in the Italian universities of the late sixteenth century, which had no conception of inertia, force, or velocity. This sort of premodern science, which was not based on the formation of verifiable hypotheses, offended Galileo's sense of how to acquire a valid understanding of nature. The other condition was the belief that human knowledge was a fixed rather than an expanding sum, most truths of natural philosophy having been ascertained by the ancients. According to this settled view, the wise researcher could do no better in solving a scientific question than to consult authoritative texts. As early as 1589, Galileo began to inveigh against Aristotelian dynamics, as he would later do at Florence against Aristotelian astronomy. His opponents thought him a seeker after specious novelties, a flashy self-advertiser. But he was a natural polemicist, with a taste for blood, and had he not been so, he would not have survived intellectually.

Until the telescopic discoveries that wholly changed his life, Galileo subsisted on a small professor's salary, yet his family expenses were onerous. He tutored students in mathematics and designed instruments, with the help of artisans and instrument-makers. At age twenty-two, he proposed an improved version of the hydrostatic or "Archimedean" balance for weighing precious metals in air and water to determine their specific gravity. After moving to Padua, he designed, at the behest of the Venetian Senate, a one-horse pump, which has never physically been recovered but was probably helical like Archimedes' famous propeller, its blades bearing the water upward. Later, in 1598, he invented a sector, a sort of compass used primarily in gunnery, and for a fee he instructed purchasers in its use. About five years later, he devised a water thermometer whereby a flask the size of an egg, with a reed-like neck marked off in arbitrary degrees, was filled with water, and depending on the local heat the water would rise to a specific degree.

Galileo's telescope is his most important invention, but his optical investigations did not end with its large-scale production. He also constructed a microscope, which led to some entomological illustration though never to extensive research, and his papers contain a "theory of the concave spherical mirror," complete with ray diagram, though he did not get around to constructing a reflecting telescope. He also designed a device to enable mariners to use the moons of Jupiter to determine longitude, and long after the Inquisition had silenced him on the subject of astronomy, when he was virtually blind, he bequeathed to humanity his second-greatest invention, the pendulum clock. It was based on his discovery of the isochronism of the pendulum—the fact (true within certain limitations) that even as the amplitude of the arc decreases over time, the period of each swing remains the same.

Galileo's writings reveal a novel ability to conjoin mathematical with aesthetic insights. Conceptually brilliant yet unacademic, he blends as only a humanist could the *esprit de géométrie* with the *esprit de finesse*. His own poetry is of the "occasional" and dilettantish variety, more mannered and piebald than what he professed to admire, but composing it surely helped to develop his fluency at prose composition. With respect to the great Italians, he regarded Ariosto as unequalled among Renaissance poets, delivered two erudite lectures on Dante's *Inferno* to the Florentine Academy, and wrote an unsparingly critical essay on Tasso's *Jerusalem Delivered*. He was alert to tropes in all their manifestations, and if he showed little patience with Tasso, this was in part because of the intrusion of allegory in Tasso's epic narrative, which he felt destroyed the unity of treatment. "Poetic fiction and fables should be taken allegorically only when no shadow of strain can be discerned in such an interpretation," he wrote. "Otherwise . . . it is like a work of art in which the perspective is forced and which, if seen from the wrong viewpoint, will appear absurd and distorted." The optical effect to which Galileo refers is called anamorphosis, and it is a historical irony, to which I will return, that this sort of distortion came to bedevil the very form of art, the frescoed dome, that the Church would increasingly favor with the aim (among others) of countering the spread of the Galilean world picture. Galileo himself is thought to have taught perspective for a while, and doubtless had complete command of the elaborate treatises on the subject that had recently been published; it was presumably on the strength of these quasi-mathematical skills that he was elected in 1613 to the Florentine Academy of Design, a glorified studio-school linked to the needs of the Medici family. His writings are studded with confident judgments on art, and he was associated with a circle of Tuscan painters, among them the aforementioned Cigoli, who were reacting against the excesses of Mannerism, which Galileo also disliked. One feels that deep down Galileo had an almost Platonic faith in graphic representation if handled with the requisite straightforwardness and homogeneity of execution: it could probe toward general, underlying structures.

It has been noted that Galileo's critical meditations on poetry, painting, and sculpture pertain to his views on the scientific method: it might even be said that they are a sort of highly colored expression of it. To a degree, this unity reflects not only his own temperament but also the mutual proximity of the arts and sciences at the end of the Renaissance. Most interestingly, Galileo did not hesitate to ventilate his fierce contempt for all composite imagery. This included pictorial intarsia, or the inlaying of wood veneers to create a picture; the Aztec featherwork then common in cabinets of curiosities; and any paintings in which some parts were not in keeping with others. He loathed Giuseppe Arcimboldo, the painter of fanciful heads assembled from vegetables, books, and other oddments. In his literary criticism, he vehemently objected to Tasso's interruption of the ongoing sweep of a lyrical tale in order to engage in some dispensable flight of fancy. What Galileo looks for in art is breadth: some form of treatment that can encompass an entire picture or story or statue with one generalizing vision and manner. The same is true of his approach to nature itself. There is no point in so-called occult or particularistic explanations of phenomena. If you have found that something is true under repeatable conditions, there is no reason not to assume that everything analogous will behave in the same way. If you have a choice, moreover, between an elaborate explanation of something and a simple one, choose the latter: not only is it more economical, but per-haps it can become the initial outline of a general principle.

There are two underlying assumptions here, so essential that we nowadays take them for granted, but they were novel at the time and exhilarating. The first is that nature is uni-form, not subject to different sets of laws at different times or places. The second, which is a kind of corollary, is that a few well-chosen experiments suffice to prove a law of nature—a scientist does not have to proceed indefinitely. Sometimes Galileo appears to work on a train of repetitive experiments, but really he is refining the same research design. If one reads the work of contemporary Galilean specialists discussing and analyzing his experiments in detail, one can follow him reformulating his law of fall or his theory of hydrostatics. As William R. Shea has pointed out, it was crucially in his experiments with floating bodies that he grew aware of the need to create conditions that could be held equal while he altered one variable at a time (a concept one could almost miss in his famous parable about the cicada's song).

Given this mindset, it was inevitable that Galileo's investigations would eventually collide with the Aristotelian conception of the universe, for the following reason if no other: Aristotle's cosmos did not recognize the uniformity of nature, positing one set of laws for the heavens and another for the sublunary sphere—that is, everything beneath the moon. The Aristotelian system is often described as a sort of onion, with the earth at the center surrounded by the three celestial spheres of the moon, Venus, and Mercury, followed by the sphere of the sun; beyond the sphere of the sun lay those of Mars, Jupiter, and Saturn, the whole being enclosed by the sphere of the fixed stars. These spheres are conceived of as circular orbits, and the heavenly orbs as perfect and incorruptible, resembling an unearthly sort of crystal. Heavenly motion is originated by divine love in the outermost sphere, and is circular because it is perfect; terrestrial motion, by contrast, is naturally rectilinear, since on earth bodies fall along a plumbline and fire rises straight upward. In point of fact, this system was emended over the centuries to account for a number of celestial observations that flagrantly contradicted it; most importantly, around

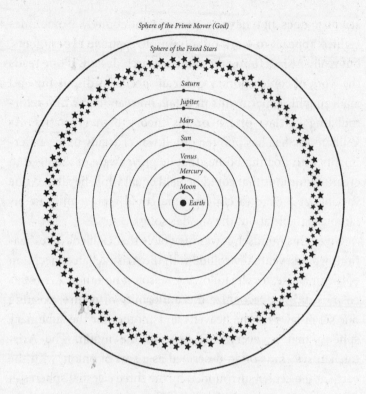

Sphere of the Prime Mover (God)

Sphere of the Fixed Stars

Saturn

Jupiter

Mars

Sun

Venus

Mercury

Moon

Earth

Simplified Diagram of the Aristotelian World System

AD 150 Ptolemy reformed Aristotle's cosmos in the vast work known as the *Almagest*. In his own writing Galileo tends casually to conflate Aristotle's and Ptolemy's systems, but since he accepts neither, he is much less concerned with demolishing their fine points than with defending the Copernican setup. Copernicus's sun-centered cosmos had already acquired many devoted followers by the 1590s, when Galileo began to embrace it, but he felt no more obliged to defend its many and elaborate computations than he did to attack Ptolemy's subtleties. What did strike him as a major scientific problem was the fact that bodies appeared to be moving in a circular fashion in the heavens, whereas they fell straight down on earth. Galileo knew that this dichotomy was merely apparent, but how could he prove it?

As noted earlier, Galileo's first observations with the telescope, conducted between 1609 and 1612, convinced him of the overwhelming likelihood that the Copernican theory of the cosmos was correct. He championed this theory explicitly if fleetingly in the *Letters on Sunspots*, of 1613, and again in *The Assayer*, of 1623. Thus it must be remembered that there is a direct logical connection between the telescopic observations, to which the central portion of this book is devoted, and the trial of 1633. He was not content, however, with three or four key observations that tended to demonstrate the validity of the heliocentric cosmos: he wanted at least the outline of a theory of celestial mechanics decisively able to refute the old Aristotelian duality and to explain planetary motion, and this theory he put into the *Dialogue*. Using the telescope, he perceived with his eyes that the vast machine we call our solar system did not function as most people had imagined for about two thousand years; in his theoretical writing, he would try to understand with mathematics more precisely how it actually did function. Yet such an attempt did not, by itself, offend the pope or the Holy Office; indeed, the pope favored the endeavor. Galileo's sin lay in his inability or implicit refusal to explain the cosmos suppositionally, as if his research were merely a fascinating mathematical exercise; because he was not toying with Copernicanism, he was passionately supporting it, deploying empirical evidence underpinned by an all-embracing geometrical vision.

It has sometimes been claimed that Galileo, though an

excellent mathematician, had no interest in advancing pure
mathematics in the sense that Descartes and Fermat did. This
argument may be true, in that the ability to work out an economical demonstration in Euclidean geometry held an abid-
ing prestige, even a sort of glamour, for Galileo and his
associates, and he did not move significantly beyond it. But
one must also remember that Galileo did not know about
algebra, though it had been used in the West for some
decades, and the mathematical culture may not have been ripe
for a great leap forward. He came close to making remarkable
contributions, however. Among his discoveries was a treatment in the *Dialogue Concerning the Two Chief World Systems*
of the acceleration of falling bodies that outlines, in prose and
with a geometrical figure, the mathematical concept of integration. Another was a discussion in the *Dialogues Concerning Two New Sciences*, the great work on mechanics that he
published in 1638, after the Church had silenced him on the
subject of astronomy, of a paradox that many students con-
front in a somewhat different form in college today. This is the
fact, in arithmetic, that there are as many squares as there are
integers, since the totality of both squares and integers is infinite. But since squares are themselves integers, it would seem
that the part, contrary to Euclid's fifth "common notion," may
be equal to the whole, in the sense of having the same number
of elements, or, conversely, that one infinity may be greater
than another. Galileo declined to investigate this paradox,
suggesting only that terms like "longer," "smaller," and "equal"
had no place (that is, required mathematical redefinition)
when applied to infinite quantities. Given his interest in infinity, he has been reproached by some historians for not pursuing the question further, as if he lived in the nineteenth

century and had recourse to set theory and some notion of the classification of sets. But it would have been hard to solve nineteenth-century problems with seventeenth-century tools.

Residing in Padua as an impecunious professor of mathematics, a subject then held in low esteem, Galileo grew attached to a young courtesan named Marina Gamba. Between 1600 and 1606 she bore him three children, Virginia, Livia Antonia, and Vincenzio, though his name does not appear on any of their baptismal records. (Marina's low station in life elicited resentment on the part of Giulia Galilei, Galileo's mother, who lived with him after his father died in 1591.) When Galileo left for Florence in 1610, he took Virginia with him, leaving the other two children in Marina's care, and in 1613 she married another man, maintaining reasonably cordial relations with Galileo. For want of dowries, both girls eventually became nuns at the convent of San Matteo; the fond letters that Virginia wrote to her father, under the name of Suor Maria Celeste, were later published and translated. She died suddenly of a fever at age thirty-three, causing him great sorrow.

Galileo's money problems were compounded by illness, which began to assail him in early middle age. He was beset by a host of complaints, of which hernias, arthritis, and bad vision appear to have been the worst, and which he complains about frequently in his correspondence. In later life he became partially, then completely, blind. His medical condition must have contributed to his often dyspeptic frame of mind. It is noteworthy that at age fifty-five he made a pilgrimage to the Santa Casa, at Loreto. The Santa Casa is the Blessed Virgin's house, which, we are told, had been borne through the air from Palestine by angels in order to escape demolition by the Turks. The usual reason to visit the Santa Casa was to pray for a miraculous medical cure, much as people do today at Lourdes. Galileo's religious faith was never in question.

It is a fact of intellectual life that people of high mathematical ability tend to feel quietly superior to everyone else. This does not necessarily mean that they are conceited or snobbish, but if challenged or obstructed by others of lesser capacity, they may abruptly show exasperation or scorn. Galileo had just such a superbly gifted mind, and as a social being he has often been put down as arrogant, irascible, and belligerent. There is ample evidence for such a view, as the list of his public quarrels is a long one. Today this may surprise us: though all these disputes are of biographical interest, only three—those with Christopher Scheiner, Orazio Grassi, and Francesco Ingoli— still prompt our intellectual curiosity, and we may fail to understand the reason for his feeling threatened or vexed by his many petty antagonists. Galileo has been said to suffer from a persecution complex; he has been called paranoid (surely a misuse of a medical term); and judging by his correspondence, he did seem to lead his life with a standing air of grievance. But here a note of caution is in order. There was no reliable intellectual property protection at this time, and a number of unscrupulous people tried to steal Galileo's ideas or to impugn his parentage of them. Many others, simply foolish or unconvinced, attacked his most profound theories or, worse, tried to get them banned. If these had been trivial notions, it would have been one thing; but they were the ideas on which modern civilization rests. In their letters, Galileo's friends, none of whom seems choleric or paranoid, often express outrage at his antagonists' envy, and they repeatedly warn him of the machinations directed against him. His life at times had a genuinely *noir*ish cast. With intermittent though mounting intensity after the Inquisition's anti-Copernican edict of 1616, Galileo's research was subject to a widely prejudicial assessment. Rarely evaluated on its scientific merits, it was more often repudiated out of hand. Just as criminals are sometimes framed by the police, or hypochondriacs fall terribly ill, so Galileo represents that not uncommon phenomenon, the temperamentally aggrieved person who actually finds himself persecuted by the authorities because he sees what they cannot or will not see.

Finally, one must take the man's social environment into account. This is an epoch in which papal legates ransack entire cities, in which the nobility bribe and blackmail one another, in which painters poison their rivals, in which cardinals commission the theft of coveted works of art. A man is ranked at least as much by the standing of his enemies as by that of his friends, and everyone has somebody he loves to insult. To insult with flair, with humor, is a fine art, and it is one at which Galileo frequently excells.

GALILEO'S GREAT ANTAGONIST at the trial of 1633, his former friend and admirer, was Urban VIII Barberini: poet, humanist, patron of the arts, and supporter of scientific research. But Maffeo Barberini was also a warrior, a fortifier of the papal domains, and staunch upholder of the Tridentine decrees on which the Counter-Reformation was based. We see him in Bernini's marble portrait of 1637–38, the face narrow, the eyes deep-set, the nose and mouth like delicate sensory instruments, as befit an aesthete. His gaze is adrift. He seems caught in the midst of conversation, listening to someone, formulating some deft reply. It is a handsome, finely articulated visage, intelligent, a trifle devious, but not especially commanding. An intellectual, not a charismatic presence.

Maffeo Barberini's pontificate lasted from 1623 until his death in 1644, and during this period he refashioned parts of Rome in the manner of a Renaissance pope. He also tried to make the political and military weight of the Papal States felt on the European stage, a task to which he proved unequal, and by the early 1630s he appeared embattled, inflexible, and given to impetuous decisions; but he was not always so. Scion of a great house of Florentine merchant princes, quick-witted, vain, cunning, and intemperate, he began his career as a scholar-priest, a master of Latin and Greek versification, a lover of nature, a serious follower of developments in art and architecture and of research at the Jesuit Collegio Romano. If he figures in the Galileo biographies mostly as a sort of ogre, a persecutor of the Tuscan scientist, his ecclesiastical biographers devote long passages to a different quarrel, and a tempestuous one at that.

When his predecessor, Paul V Borghese, opted for Carlo Maderno's redesign of Saint Peter's, which proposed to elongate the nave at the expense of Michelangelo's façade, Maffeo Barberini, then a cardinal, bitterly opposed the desecration. It was about as close to fisticuffs as the college of cardinals ever came. To his face, Maffeo told Paul V that should he eliminate Michelangelo's design, the next pope would tear down the new structure and put everything back to rights, at which Paul V shot back that it would be built to last forever. The next pope was Gregory XV Ludovisi, who was old and ailing and lasted only a few years; then came Maffeo himself, and though he never managed to replace Maderno's rather tame façade with Michelangelo's original and vigorous one, much else in

"Michelangelo presents his model of St. Peter's to Pope Julius II,"
by Domenico Cresti da Passignano

Passignano, a colleague of Cigoli, is credited with being the first to observe the
sunspots through a telescope; Cigoli forwarded their discoveries to Galileo. In this picture Passignano implicitly criticizes the papacy for its failure to realize Michelangelo's
design, a decision of Pope Paul V's that also infuriated Urban VIII. Urban also reportedly disagreed with Paul's promulgation of the 1616 edict against the teaching of
Copernicanism, yet he later persecuted Galileo for infringing it.

his early career confirmed his desire to make the Church a sponsor of the finest in the arts and sciences.

Maffeo Barberini wrote books of poetry in both Italian and Latin, which together were reprinted more than twenty times during his pontificate. (Bernini illustrated an edition of 1631.) Maffeo's poetry, like much conventional verse of the period, evokes the moods of nature, the passage of time, the lives of the saints, and the intimations of mortality, but it is also bedeviled by a Baroque sense of theatrical illusion so intense as to become a metaphor for universal vanity. We are all understudies for ourselves, he seems to suggest; to become one with our mortal roles would be to show insufferable pride. Maffeo nursed para-doxical emotions toward creativity, his own included, as though his admiration for human achievement were adulterated by an almost toxic sense of futility: a certain forlornness lurked behind that mask of robust ambition. One of his pet projects was his own tomb, faced in many-colored marbles, which Bernini built slowly during his lifetime in Saint Peter's.

Urban VIII Barberini is generally regarded as one of the most shamelessly nepotistic of pontiffs. No sooner had he been elected to the office than he began to place his relations in positions of power. Within six weeks his nephew Francesco, an intelligent twenty-six-year-old who later enjoyed Galileo's friendship, became a cardinal. The position of "pope's nephew" was a formal title, universally honored; for ages every pope had had a "nephew," sometimes not genuine kin, who served as a virtual secretary of state. Yet temperamentally Francesco was poles apart from Maffeo. He had an apartment in the Vatican and handled much of the pope's diplomatic agenda, but he did so with unusual circumspection, as though submitting with resignation to inclement weather. The pope's brother Antonio and a nephew, Taddeo, aged only nineteen, soon followed Francesco into the curia.

Maffeo also exploited his position as a source of lucre. Ecclesiastical posts not granted to family members were simply auctioned off; the distinction between the pope's personal purse and the Vatican treasury grew progressively fictive; and the Barberinis exerted their influence to acquire the palaces or landed estates of several great Roman families who found themselves short of funds. It is necessary, however, as with Galileo, to see such behavior in context. Early modern Italy (like much of southern Italy today) was dominated by an ethic whereby family ties trumped ordinary ethical considerations. No man could possibly rise to a position of power without indebtedness to a number of people in his clan, and rewarding them became a paramount obligation. The Barberinis were hardly alone in acting as they did; had they not been ruthless, like the Orsinis and the Borgheses and the Colonnas and the other great families of the period, they would have met the fate of the Gonzagas, in Mantua, who let their guard down and saw their city sacked and their palace looted a few years later. The ceiling of one of the chambers of the duke of Gonzaga bore a gilded design of a labyrinth with the phrase *Forse che sí, forse che no* running repeatedly in all directions. It meant "Maybe yes, maybe no," like a memento mori, and as Maffeo Barberini knew, the only measure likely to gain one a positive fate in this cloak-and-dagger world was surrounding oneself with blood kin and building a mighty war-chest. Not one to take chances, Maffeo made a point of fortifying Rome and the entire papal principality. Yet even by the standards of their day the Barberinis were regarded as excessively greedy. As a popular pasquinade had it, "What the barbarians didn't steal, the Barberinis did."

Urban VIII's art and architectural patronage served multiple functions. It satisfied his excellent taste, glorified himself and his dynasty, and maintained the prestige and grandeur of Rome in the eyes of foreign envoys, Catholic prelates, and pilgrims, who brought in cash. Both as a bishop and a cardinal, Maffeo acquired splendid pictures by Raphael, Andrea del Sarto, Caravaggio, and many others. As a pope he also undertook the decoration of the interior of Saint Peter's, the construction of a family chapel in Sant'Andrea della Valle, and the rebuilding of the church of Santa Bibiana, a matter of little account had it not been entrusted to Bernini.

Gian Lorenzo Bernini, who lived from 1598 to 1680, was recognized from childhood as a genius and largely defined the style we now call the Baroque. Maffeo had long yearned to gain Bernini's services for himself. At first he lacked the means to do so, but as pontiff he virtually monopolized Bernini and so became the greatest patron of sculpture Rome had ever known. Within a year of his election to the papacy, Urban VIII placed Bernini in charge of his entire artistic program, enabling the sculptor-architect, then still in his twenties, to begin to refashion the visual impact of the city. Bernini also became his intimate friend, closer to him than any councilor or member of the Barberini clan. The story is told, and there is no reason to doubt it, that when Bernini was sculpting his *David*, for whose face he used his own likeness, Maffeo would hold the mirror for him, an act of sublime (if not almost servile) admiration. Often in the evenings Maffeo would welcome Bernini into his apartment in the Vatican, where the pair would discuss their vision of Rome until the pope drifted into slumber. Then Bernini would quietly close the windows, draw the curtains, and depart.

We must look closely at Bernini because his early work is the concrete embodiment of Maffeo's own sensibility. Yet when we examine this connection, we discover a bewildering irony—that Maffeo would probably have been delighted to sponsor not only Bernini but also Galileo. Of course, such a development was impossible—Galileo was attached, at a high price, to the court of Tuscany—but Maffeo was extraordinarily well disposed toward Galileo and kept close track of his work. As early as October in 1611, Maffeo, then a cardinal, expressed support for Galileo during a scientific debate over the nature of flotation at the Pitti Palace, in Florence. In May of the following year, upon receiving Galileo's treatise on the subject, he wrote him a letter praising his "rare intellect" and suggesting that their minds vibrated in harmony. In 1620, Maffeo wrote a poem praising Galileo—inevitably, it mentioned the moon—and in 1624, as newly elected pope, he granted Galileo's son, Vincenzio, an annuity of sixty crowns and urged Galileo to write the definitive treatise on Copernicanism, a work that later became the *Dialogue*. We will see exactly how the pope came to prosecute Galileo for writing the very book that he himself had proposed, but Galileo was by no means the only adventurous thinker whom Maffeo, a late child of the Renaissance, encouraged and supported. In 1626, he rescued Tommaso Campanella from the dungeons of the Inquisition and welcomed him into his intellectual circle: Campanella, one of the most audacious philosophers of the period, had defended Galileo and posited an infinity of worlds.

I do not wish to suggest that Bernini and Galileo were in some sense conceptual counterparts, but rather that Galileo appealed to Maffeo Barberini in much the same way that Bernini did, though mostly from a tantalizing distance. Dur-

ing the Baroque period, artists and scientists did not seem as
different as they do today. Not only was Galileo himself an
accomplished musician, an excellent prose stylist at his best, a
sometime poet and literary critic, and a friend of some brilliant artists, but he also sought a certain elegance in his geometrical proofs and in his increasingly public pedagogy.
Bernini had engineering as well as artistic skills, and there was
a speculative component to his art. For the entire Baroque
enterprise, representing space and creating shapes meant
acquiring greater knowledge about statics and dynamics,
light, cast shadows, and the science of perspective: it meant
reaching out and grasping space. Just as Galileo took a fixed
system of the cosmos and tried to show how the earthbound
observer actually revolved within it, Baroque artists like
Bernini took a frontal, cinquecento vision of order and
twisted it into curving and spiraling planes.

Bernini's vision was too vast and complex to be easily
summed up. If one looks, however, at a sculpture like *Saint
Longinus* in Saint Peter's, which he worked on for Urban VIII
before and after the years of the Galileo trial (1629–38), or at
the draperies on the angels for the Sant'Angelo Bridge span-
ning the Tiber, which came decades later but represent an elab-
oration of the same idea, one notes right away an original
conception of gravity—and gravity of course was one of
Galileo's chief concerns. In the paintings and sculpture of the
High Renaissance, which Bernini wished not to subvert but to
extend by novel means, the effect of gravity upon fabric is generally conveyed with simple decorum. For instance, in
Raphael's Tapestry Cartoons, the great studies he and his assis-
tants executed in 1515–16 as designs for the Sistine Chapel
wall hangings, the robes of the biblical characters appear by

and large to be woven of plain, undifferentiated wool and to fall toward the ground in broad folds, without frequent hooking or complicated tubular effects. Curiously enough, if one thinks about it from the astronomical point of view, in one way or another these draperies fall toward a point at the center of the earth. They are subject to gravity, which is, after all, part of our mortal dispensation—unlike the angels, we cannot fly. But they are subject to gravity in a particularly graceful way.

With Bernini's drapery something else happens. Even though his *Longinus* stands still, and no wind billows his robe, he seems to be swept by tremendous energy. His garment—fluttering, tangled, and wavy—does not accentuate the saint's gesture but enacts a continuous independent occurrence, or a captured moment within such an occurrence, for in fact no fabric ever hangs this way—it is an antigravity effect. Bernini sees grace, in the sense of something freely, divinely given, as a medium resistant to gravity, permitting a form of existence alien to our own—potentially, many of Bernini's figures could lift off, like helium balloons—and as one looks up into some of his domes one sometimes does see *putti* playfully levitating over the cornices. In a way, this medium of grace is the super-oxygen of the Counter-Reformation, in which street procession and theater and miracle can breathe. And it is further developed in Bernini's architecture and that of his colleagues.

The Baroque style does many things, but one of them is to take the rather static classical architectural vocabulary, subject it to curvature, and rotate it at will through at least thirty degrees in plane and in elevation. If there is no strict correspondence between Bernini and Galileo, who worked in very different fields, they do share a certain mental attitude and a dynamic grasp of space. They have a common willingness to contem-

plate what might at first seem implausible solutions, a fascination with the time factor—with the combination of direction
and velocity—a keen interest in curvature and curved planes,
and a constant perception of experiential relativity. The
Baroque style attempts to put High Renaissance plastic values
into motion, to capture the semblance of an ever-shifting world.
Stage sets implicitly rotate the audience off-center, so its members observe the action from a variety of angles. Architectural
elements are enjoyable or interesting much less in themselves
than as parts of a whole perceived by a circulating observer. The
sense of reality becomes dependent on viewpoint and lighting.
In a painting, for example, if a shadow causes the limbs of two
figures to disolve optically into each other, no contour will be
provided to establish their independent identities. Though
Bernini's art was an expressive, not a scientific, enterprise, and
certainly in no sense Copernican, it resembles the heliocentric
model in its emphasis on relativism and on the observer's position as opposed to the fixed contemplation of forms.

Bernini's baldachin, an ornamental canopy in Saint Peter's
which looks different from every angle, or his Scala Regia in the
Vatican, a long staircase with ever-changing and eye-deceiving
widths, not to mention his fictive seating areas and imaginary
audiences, all depend for their impact on the spectator's motion
through space. Bernini did not invent the Baroque style, but his
work expanded the optical Rome that had begun under Sixtus V
with Michelangelo's Campidoglio and continued radically
throughout the 1590s. The old Rome had been a hodgepodge of
evocative structures in which the pilgrim struggled to find an
axis of symmetry whereby he could orient himself; basically, he
always had to ask his way around. The new, optical Rome
offered landmarks: it was characterized by straight thorough-

fares with an obelisk centered at the focal point, on the axis of symmetry of an important façade, like a church. The visual channel of the avenue thus worked analogously to a telescope or gun sight (though without the element of magnification).* When one entered the church, one looked up into a dome like a spiritual planetarium, a model of heaven—the upper half of what astronomers called the "celestial sphere"—with, at the zenith, a glazed lantern as the focus. (In some cases, like Sant'-Andrea della Valle, the dome's interior surface bore an actual picture of heaven.)

This was the Rome that Maffeo Barberini strove mightily to renew and to expand. He couldn't help being fascinated by both Bernini and Galileo because he was so captivated by the Baroque idea of space—by his contemporaries' power to create a semblance, whether in stone or in mathematical theory, of a world that wasn't static but that changed unceasingly, like the one we inhabit. It was natural for the pope who backed Bernini to encourage Galileo, and his disappointment was natural later on when Galileo overstepped (as Maffeo saw it) the Church's clearly stated guidelines for the formation of scientific hypotheses. Unfortunately, Maffeo was more than disappointed, he was outraged, and his outrage caused so much damage that the Church has not wholly repaired it to this day.

THE GALILEO TRIAL has been interpreted numerous times, from various ideological standpoints. It is clearly the prime

---

* Anticipating the invention of the telescope, Michelangelo's Campidoglio, in Rome, begun in 1538, was also designed so as to appear magnified to the approaching observer, by means of scenographic perspectives. Its construction was delayed until after Galileo's death.

drama in the history of the conflict between science and religion, more central than the evolution controversy of the 1860s, which never formally pitted Darwin against the Church of England, and certainly more consequential than the rather jocular Scopes trial of 1925. In Galileo and Urban VIII's day, Rome was the capital of a sovereign theocratic power, the Papal States, which in 1600 had had Giordano Bruno burned at the stake for refusing to abjure his heterodox philosophy, and which reserved the right to torture (and, ultimately, to execute) Galileo, should he appear to withhold evidence. The trial also marks a turning point in the evolution of freedom of thought, since the great Tuscan physicist had been ordered not to pursue his research but had done so anyway, and had published the results. As if this were not enough, still more was at stake, for Galileo's accusers were alarmed not only by what he said but also by how he said it—by his reinvention of certain components of the scientific method, neglected in the West (though not in the Islamic world) since antiquity. Clerical scrutiny of his writings indicated that they appeared to assign logical priority to empirical observation over Scripture, which was contrary to Catholic doctrine.

The truth is that Galileo had been running a terrible risk since 1616. Did he know this? Perhaps not; perhaps he didn't much care, until the men in black vestments began threatening his freedom. At the time, the coming science-religion clash was not foreseeable, and it remains puzzling today for reasons partly unexplored, though clearly the leaders of the Catholic Reform saw the scientific revolution as forming an analogy with the rise of Protestantism, a terrible mistake. To a degree easily forgotten, dynastic rivalries and personal emotions also came into play. Historians of science have generally discussed the debate between Galileo and the Church as though it were *cosa mentale*, the way painting was for Leonardo da Vinci. But the trial was not a mental thing, like a modern scientific colloquium, but a series of frightening interrogatories held before a tribunal of the Inquisition, and each side's perception of the other was critically defective. The Roman cardinals respected Galileo but had their own peculiar anxieties, exacerbated by the spread of heresy, by dynastic and geopolitical rivalries, and by internal, ecclesiastical rough-housing; nor did Galileo's argumentative personal style make his cause any the easier.

From the modern Anglo-Saxon standpoint, Galileo's trial bears little juridical resemblance to anything we would call by that name. It is far more like the informal hearings held by government, corporate, or religious bodies that do not require the observance of strict rules of evidence or the protection of the rights of the accused but might result all the same in his or her professional disgrace. A huge literature exists about this "trial," but almost none of it attends to the ragged formlessness of the proceedings, except perhaps to note their deviations from canon law. As a rule, such writing explores the conflict between Galileo's ideas and those of the Church, and this is understandable, because no abrasion is more central to our age than that between science and religion. Intellectually, this is where the interest lies—or ought to lie. In fact, however, the Church scarcely contested Galileo's science, but only enjoined him against pursuing it freely, with the consequence that the "debate" between Galileo and the Church is merely a mental construct, a dichotomy posited to satisfy a hunger for symmetry. It is not as if Galileo and Pope Urban VIII ever went, as we say, "nose to nose." In fact, it was the Aristotelian intellectual establishment that vociferously opposed Galileo; the Church's substantive arguments against his methodology were assembled later, as an afterthought, when he had already been silenced. If there is, then, a kind of intellectual weightlessness to this nondialogue, Galileo's emotional predicament acquires thematic tension between September of 1632 and June of 1633, as a result of the encounter between two distinct modes of reasoning, feeling, and wanting. The scientist thought in one way, the Church in another; they operated on different psychic planes, a condition most clearly perceived by the Tuscan ambassador to the Holy See, Francesco Niccolini. The question we may wish to ask ourselves, not now but at the end of our story, is whether this great trial, so pregnant with consequences, does not reveal certain constants in the debate between science and religion.

2

The Telescope; or, Seeing

Until 1609, when Galileo trained his telescope on the sky, the proponents of the Copernican system had in a sense been navigating blind. They had a map that was essentially correct, but no visual grasp of the world in which they were traveling. In a flash the telescope changed that. Now Galileo and his colleagues could see. They knew they were on the right track. But many others could not see, or refused to see. For this sort of outmoded savant, Johannes Kepler, writing to Galileo on March 28, 1611, had some interesting words in Latin, the scientific lingua franca of the day. Such a person was stuck in a *"mundo chartaceo,"* a world of paper, Kepler said. *"Negatque solem lucere"*—it was as if he wouldn't admit the light of the sun. *"Caecus"*—he was blind. Not by force of circumstance, but of his own foolish will.

STILLMAN DRAKE HAS claimed that Galileo had concluded as early as 1595, the date of the publication of *De motu*, a treatise on mechanics he wrote while still at the University of Pisa, that as a geometrical construct the Copernican hypothesis came closer to explaining the motions of the celestial bodies than the alternative, Ptolemaic or geocentric system. This may be so. It is certainly true that Galileo, in a letter of 1597 to Jacopo Mazzoni, a friend and ex-colleague at Pisa, offhand-edly assumed the earth's motion in a proof concerning the vis-ibility of the stellar sphere. Indubitably his evolving conception of mechanics clashed with the geocentric cosmol-ogy, which he saw as dependent on outdated Aristotelian con-cepts of gravity, force, and motion.

At this period, astronomy was regarded as an exploration of mathematical suppositions, in part for complicated metaphys-ical reasons and in part because mankind had very scant empirical knowledge of the heavens. Observers going at least as far back as Hipparchus had made clever calculations and predictions, but they had no way of scrutinizing celestial bod-ies. This notion of reasoning about the heavens "supposition-ally," *ex suppositione*, mind-bending to us today, is rooted in a certain premodern conception of a hypothesis not as a theory likely to be empirically confirmed, or already largely con-firmed, but as kind of a logical holding pattern. Most astron-omy before the invention of the telescope fell into this category. Data would be gathered and a mathematical system would be mapped into it, which approximately fit the facts ("saved the appearances," in the language of astronomers). These planetary models were required to fit the constraints of Aristotelian physics, but could not, beyond that point, be empirically verified.

In 1597, Galileo wrote Kepler that he had already accepted, on the basis of mathematical physics, both the rotation of the earth and its revolution about the sun. However, he was unsure of his ideas and loath to publish a paper on the subject. But the appearance in the autumn of 1604 of a nova, an explosive variable star whose luminosity increased by many magnitudes during a period of hours, prompted him to give three lectures at the University of Padua, each attended by more than a thousand people, in which he noted the absence of parallax for the nova, that is, its apparent motionlessness with respect to other celestial bodies. He correctly placed it among the stars, precisely where Aristotelian doctrine asserted that no change ever took place. Though he was not overtly espousing Copernicanism, he was bluntly attacking the alternative theory, and the audience was left to draw its own conclusions.

So far Galileo's Copernicanism had been a mathematical hypothesis. What transformed it into something resembling a modern scientific theory was his radical improvement and use of a gadget recently devised in northern Europe. This was a tube with a weak lens at each end, which Galileo almost at once converted into a precision instrument. When we consider this extraordinary invention, we may wonder what a person living in 1609 would have needed to perform such a feat, and the answer is rather simple. He would have needed to know, or figure out, something about geometrical optics. He would have needed to possess a finely tuned lathe. And he would have needed to have access to superior glass and to abrasives for lens-polishing.

Geometrical optics is the science of how light behaves in an optical system, such as a telescope, where its wavelength may be regarded as negligible compared to the dimensions of the lenses or other components. We now learn in grade school that light is a wave phenomenon moving at a speed of 3×10^{10} meters per second in empty space, and that when light enters a medium such as glass its speed depends on the density of the medium and the wavelength, or color, of the light. It is also refracted, or bent, because when a light beam progresses from a less dense into a denser medium it bends toward "the normal," the plane of the medium it encounters. Light enters the medium at an angle to the normal of the surface, called the angle of incidence, creating the beam's new angle within the glass, called the angle of refraction. Snell's law of refraction is a simple trigonometric formula that tells us the index of refraction of optical glass—how beams of light bend as they pass through lenses and thus how any lens creates the image that it does. But this formula was worked out only in 1621 by Snell (Willebrord Snellius), who did not publish it, and again, independently, by Descartes, who did publish it in 1637, long after Galileo constructed his telescope. So though Galileo knew something about refraction, which had been studied since antiquity, he could not identify the exact optical properties of the lenses he was starting to make.

Glass itself is manufactured by bringing silica plus other components to liquidity and then cooling them. Venetian glass was made of quartz pebbles composed mostly of silica, which was milled into sand and combined with soda ash imported from the Levant. Optical glass was presumably discovered by accident—perhaps because it occurred in bull's-eye windowpanes or the bottoms of bottles sometime in the thirteenth century, and by about 1260 spectacles of the sort that we call reading glasses could be bought in Florentine shops. Yet strangely, no optical instruments, no microscopes or telescopes, were produced for the purpose of getting a closer look at nature, and about eighty years ago an Italian historian of optics named Vasco Ronchi suggested a reason for this. A sort of conspiracy of silence, he said, had been mounted against the research potential of the lens. The pre-dominant medieval theory of vision, derived from an ancient Greek idea, held that invisible films, called *species*, issued from the eyes and assembled optical data, so that anything impeding their progress, such as mirrors, prisms, or lenses, deformed and corrupted these images. Sight as such seemed an unreli-able sense, making objects appear smaller and grayer as they receded toward the horizon, so to alter sight still further with lenses, all of which had intrinsic distorting properties, would scarcely have improved matters. Scant documentation exists to support Ronchi's thesis, and it was vigorously disputed in 1972 by David C. Lindberg, a highly knowledgeable American his-torian of optics, who pointed out the scope and value of the medieval work in this field. It can be argued that no prejudice against lenticular devices as a research tool existed, but that it simply occurred to no one to devise such things; or, alterna-tively, that artisans may have tried to concoct a spyglass and failed, and being illiterate, left no record of the attempt. In any event, we do know how the first one was invented.

A primitive telescopic effect is attained when two lenses of suitable focal length are aligned on the optic axis between the object to be viewed and the retina of the viewer and a clear, enlarged image is perceived. Since the first people to jerry-rig telescopes probably did so to amuse themselves, pulling apart spectacles and lining up lenses, one wonders what corrective purpose the original lenses served and what they looked like. Most people buying spectacles, then as now, were probably constant readers over age forty—clerics, scribes, scholars, lawyers. The overwhelming popular demand would have been to compensate for far-sightedness, or presbyopia (the eyes' progressive inability to accommodate to near points) and only

later for near-sightedness, or myopia, which is less common.
The earliest lenses, of whatever sort, had a faulty spherical
curvature—that is, the glass piece was a sector of a sphere—
and a focal length of 12 to 20 inches, the focal length being the
distance at which an object viewed appears in focus in the lens.
(In the measurements of today's drugstore reading glasses, that
would be about 2 to 3 diopters, a diopter being the refracting
power of a lens expressed in meters and in reciprocals: thus a
lens of focal length 20 centimeters has a refracting power of
1/0.2 meters, which equals 5 diopters). They were made of glass
blanks, which were ground on a lathe attached to a bench.
Convex lenses helped correct presbyopia, concave lenses
myopia, and one might assume that the concave came later,
being harder to grind and harder to adjust to the customer,
who might need to be fitted on a trial-and-error basis.

When did the first primitive spyglasses begin to appear? In
1578, William Bourne, in a book titled *Inventions or Devises*,
mentions that "to see any small thing a great distance from
you, it requireth the ayde of two glasses," a puzzling phrase
that might refer to a telescope designed by Thomas Digges, an
Elizabethan astronomer whose observations of the supernova
of 1572 were consulted by the great Danish astronomer Tycho
Brahe. Bourne, however, noted that the tiny field of view covered by this gadget was a great "impediment" to its use
(though this would also be true of Galileo's telescope). The
instrument, on the assumption that it actually existed, might
have been made with very impure glass.

The first telescope that indisputably worked was offered in
September of 1608 by a spectacle-maker named Hans Lipper-
hey to the States-General of the United Provinces, in The
Hague. Lipperhey was a Westphalian who had moved to Mid-

delburg in Holland, and he wanted a patent. Instead he was requested to make a pair of binoculars, presumably for military purposes, and to use quartz instead of glass; within a few months, he gave the Dutch government three binoculars and was handsomely rewarded but received no patent. The reason, as Albert Van Helden has discovered, was simple. A Zeeland government document shows that a "young man" had meanwhile come forward "with a similar instrument," and "we believe that there are others as well, and that the art cannot remain secret." In fact there seems to have been considerable experimentation with such optical instruments in the glass-making town of Middelburg, where several Italians worked, presumably introducing additional technical information from their homeland.

By the summer of 1609, the Dutch telescope in one form or another had been around for about nine months. It had been sold to the king of France and his prime minister, and to the archduke of Austria, who governed the portion of the Low Countries under Spanish rule; it had been shown at the Frankfurt autumn fair of 1608; and it had been obtained by the papal nuncio to Brussels, who sent it to Cardinal Scipione Borghese in Rome, an unscrupulous prelate of excellent taste and wide influence, from whose hands it had proceeded to the pope's. That July an English mathematician-astronomer, Thomas Harriot, working out of Syon House, the seat of the earl of Northumberland, used a telescope with the power of x6.3 to view the sunspots and Jupiter's moons. In so doing he preceded Galileo by about five months but with—crucially— about five times less magnification. Spyglasses arrived in Milan by May, in Rome and Naples by July, and in Venice and in Padua by early August.

Later recalling his dramatic improvement of the instrument, Galileo wrote that

> a report reached my ears that a spyglass had been made by a certain Fleming by means of which visible objects, although far removed from the eye of the observer, were distinctly observed as though nearby. About this truly wonderful effect some accounts were spread abroad, to which some gave credence while others denied them. A few days later the rumor was confirmed to me in a letter from the noble Frenchman in Paris Jacques Badovere, which caused me to apply myself totally to investigating the means by which I might arrive at the invention of a similar instrument. This I did shortly afterward on the basis of the science of refraction. And first I prepared a lead tube in whose ends I fitted two glasses, both plane on one side while the other side of one was spherically convex and of the other concave. Then, placing my eye near the concave glass, I saw objects satisfactorily large and near, for they appeared three times closer and nine times larger than when observed with the naked eye alone. Next I constructed another one, more accurate, which represented objects enlarged more than sixty times [that is, what we would call x8+]. Finally, sparing neither labor or expense, I succeeded in constructing for myself so excellent an instrument that objects seen by means of it appeared about a thousand times larger and more than thirty times closer [that is, x30+] than when regarded with our natural vision.

The truth was that in early August a Dutchman had arrived in Galileo's own town of Padua and was showing his spyglass to potential purchasers. So before Galileo had much of a chance to improve his first telescopes—though he probably acquired better glass than whatever his rivals were using and was deepening the sphericity of the eyepiece—he decided that he had to design a superior instrument and stake out a market. Eight years later, Galileo recalled this key moment a little differently than in the foregoing description, hinting obliquely at the ferocity of the competition and mentioning a Dutchman active in Germany, though not his rival in the Veneto. "News came," he wrote in this alternate account of his discovery, "that a Hollander had presented to Count Maurice [of Nassau] a glass by means of which distant things might be seen as perfectly as if they were quite close. That was all. Upon hearing this news, I returned to Padua, where I then resided, and set myself to thinking about the problem. The first night after my return, I solved it, and the following day I constructed the instrument and . . . afterwards I applied myself to the construction of a better one, which I took to Venice six days later." This version, which omits to mention the letter from Badovere, stresses the rapidity of Galileo's improvement of the spyglass, and indeed he got to work with extraordinary speed. He needed to, for there was competition not only from the itinerant Dutchmen but from very knowledgeable astronomers. In England, as mentioned earlier, Thomas Harriot had already mapped the moon in August, using his weak telescope, and on the Continent four mathematicians, including Father Christopher Scheiner, were already in possession of telescopes.

Galileo was thinking fast. "My reasoning," he wrote, "was this. The device needs either a single glass or more than one. It cannot consist of one alone, because the shape of that one would have to be convex . . . or concave . . . But the concave diminishes [visible objects]; and the convex, while it does indeed increase them, shows them very indistinctly and confusedly. Therefore, a single glass is not sufficient to produce the effect. . . . Hence I was restricted to trying to discover what would be done by a combination of the convex and the concave. You see how this discovery gave me what I sought."

Joining a bitter race against the Dutchman who was nearing Venice, Galileo wrote on August 4 to his friend the savant Fra Paolo Sarpi, imploring his help. Obligingly, Sarpi, who had great influence with the Venetian Signoria, asked to inspect the Dutchman's spyglass and advised the Signoria against its purchase. On August 21, a breathless Galileo arrived in Venice with a more powerful instrument than any then in use—its magnification was about x9—which he proudly proceeded to demonstrate to a group of Venetian senators and naval officers from atop the campanile in the Piazza San Marco. To their delight, they were able to sweep a radius of some fifty miles with the instrument, even beholding the faithful entering the church of San Giacomo on the glassmaking isle of Murano. On August 24, 1609, Galileo wrote to the doge, Leonardo Donato:

> Most Serene Prince, Galileo Galilei, most humble servant of Your Serenity, assiduously laboring by night and with unwavering spirit to satisfy not only the duties of lecturer in mathematics in the University of Padua, but with the hope of finding some useful and notable discovery to bring extraordinary benefit to Your Serenity, now appears before the aforesaid with a new device, a hollow spyglass of the most unusual perspective specifications, in that it conducts visible objects so near to the eye and represents them so large and distinct, that something which is distant, let us

say, by nine miles, appears to us as if only one mile away: an invaluable thing for any maritime or terrestrial enterprise, being able at sea, at a much greater distance than usual, to descry the masts and sails of the enemy.

In turning to the Venetian Republic, his employer and a great naval power, Galileo followed the inventor's time-honored tradition of first approaching the military to raise money. The plan worked: the Venetian navy bought his tele-scope, the Senate doubled his salary at Padua and renewed his position for life, and he spent the fall of 1609 trying to improve the instrument's magnification. The alacrity with which the already forty-five-year-old Galileo seized on this optical idea, the speed with which he perfected it with his own hands, the confidence with which he presented it to the Republic, and the fortitude with which he confronted his rivals—all distinguish him as a man of action as well as an intellectual.

Almost at once, Galileo's detractors began to belittle his role in the development of the telescope. But Galileo liked to argue that his knowledge of earlier prototypes made his task as inventor not easier but more difficult. The original inventor, he claimed, had merely stumbled on the device: he himself, on learning of the concept, had to see his way clearly forward to something of scientific utility. He did not admit to having handled any of the playthings then popular in certain cities, and perhaps he had not done so by the time he set to work on his first version: it scarcely mattered, since a drawing or verbal description would have sufficed to acquaint him with the basic idea. What is certain is that he had greatly outdistanced the available prototypes within a matter of weeks.

Now, despite what Galileo said about his knowledge of the

"science of refraction," it appears that only Kepler had a good enough mathematical understanding of geometrical optics at this time to trace satisfactory ray diagrams. Galileo's advantage lay rather in his deep faith in the scientific relevance of seeing, of vision. Though he worked from the start with two lenses, let us pretend, in order to better understand Galilean optics, that he first examined one lens only, affixed to a tube. Holding a plano-convex lens—a lens spherically convex on the side facing the object to be viewed and flat on the other—up to an object, let us say the moon, he could theoretically focus his eye on a tiny, inverted image of the moon suspended upside-down in space, somewhat magnified, in the plane called the prime focus of the lens. In effect, the lens had converted the arriving rays of light into cones whose apexes fell approximately in this plane. Brief study of the adjoining ray diagram will disclose why the image is inverted. (If the image is captured on a piece of paper, as in the rudimentary camera obscura later devised by Galileo's disciple Benedetto Castelli and also by his friend the painter-architect Ludovico Cigoli with the help of Domenico Passignano, using two lenses positioned to throw the focal plane well to the left of the telescope, greater image stability and a wider angle may be achieved, with some loss of brilliance.)

We know, however, that Galileo was working with two lenses from the outset. We may imagine him experimenting with various pairs. And it is indeed possible, if you have a selection of weak lenses, to run through the effects produced by every possible combination of one concave and one convex lens with respect to magnification and image sharpness. You hold them in front of your eye, one close, like a loupe, the second one behind it, and gradually extend the second one out to see what happens. The exercise is tedious, and Galileo didn't

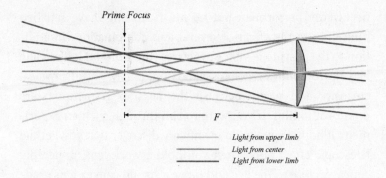

Ray Diagram of a Telescope without an Eyepiece

The focus is on a remote light source (like the moon) to the right. The location of the apexes of the bundles of light on the plane of the prime focus shows that this (aerial) image is inverted.

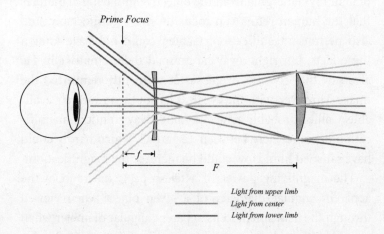

Ray Diagram of a Telescope with a Negative (Concave) Eyepiece

The eyepiece both magnifies the image and converts the bundles of light transmitted by the objective into parallel bundles more easily processed by the eye. As noted in the text, the angle at which the rays enter the eyepiece and that at which they leave it diverge, and this factor is represented by the ratio *F/f*.

necessarily perform it, but he must certainly have satisfied himself as to the effect of several lens combinations in addition to the "bagatelle"-like spyglasses (the word is Galileo's) he had heard about or got his hands on, partly out of physical curiosity and partly because the laws of refraction were unknown to him. If you were to run through such an experiment at home, you would eventually discover that you get the telescopic effect only when you hold a weak lens, preferably concave, close to the eye and move a strong convex lens gradually away, until the focal lengths of the two coincide at the point of their greatest magnification and clarity. So, having presumably satisfied himself regarding the effect produced by a single lens, Galileo began to rig up test spyglasses with two lenses. The concave eyepiece converts the converging bundles of light rays emerging from the objective into parallel bundles that the human retina can focus on more comfortably, and also permits magnification: Galileo could not have known these facts, but right away he grasped them empirically. He also saw that concave eyepieces conveniently reinverted the image. However, with the eyeglass lenses then routinely available, Galileo probably achieved in this way an initial magnification of somewhere between x2 and x3, which hardly could have satisfied him. How could he increase the magnification?

The magnifying power of a telescope is the ratio of the apparent angular diameter of a given object when viewed through the instrument to its apparent angular diameter when viewed by the naked eye. In other words, if you can find a way of measuring the angle subtended by the diameter of the object—the moon, say—when seen through a telescope and compare it to the angle subtended by the moon when seen by the unaided eye, then you know the power of your telescope.

This could be confusing back then, though nowadays, by combining Snell's law and an equation called the thin-lens formula, which allows us to regard the glass in question as a medium of negligible thickness, you can easily determine the power of a telescope like Galileo's: its magnifying power equals the ratio of the focal length of the objective to the focal length of the eyepiece. In other words, by dividing the former into the latter, you can learn the power of your telescope. In experimenting with the focal length of his objectives, Galileo must soon have had some rough notion of the equation $M = f_o/f_e$, where M is power, f_o is the focal length of the objective, and f_e the focal length of the eyepiece, because working in his own shop he soon ground objective lenses of ever-longer focal lengths.

It seems that so far very few others had been able to intuit the relation between magnification and the focal lengths of the two lenses, the eyepiece and the objective. Why did Galileo advance so rapidly where others failed to progress?

Most of Galileo's early competitors were lens-grinders. And magnification is a simple function—inverse proportion—that might elude a lens-grinder but be rapidly divined by a mathematician. Yet since there is no documentary evidence that Galileo figured out the formula for magnification, it is more likely that he came to understand the primacy of objectives with long focal lengths through trial and error. After all, he was good with his hands and interested in machines of any sort. The problem of constructing a telescope had at once fascinated him, as had that of the helical pump and the gunnery sector.

When Galileo was a very old man, in June of 1640, he wrote a letter from Arcetri to Fortunio Liceti, a professor at the University of Bologna, in which he lamented "having always been in darkness concerning the essence of light." In August of the same year, he again wrote Liceti that "I have always considered myself unable to understand what *lumen* [light] was, so much so that I would readily have agreed to spend the rest of my life in prison with only bread and water if only I could have been sure of reaching the understanding that seems so hopeless to me." This confession, in which we hear the accent of the true scientist, has been tied to the brief time that Galileo spent in the study of geometrical optics, especially considering that Kepler's breakthrough treatise on the subject, *Ad Vitellionem Paralipomena*, had already appeared, in 1604. But actually it seems to reflect Galileo's despair of understanding not geometrical optics but something deeper, the physical nature of light itself, which he puzzled over throughout his career.

Galileo discussed magnification very warily, never offering his readers more than a simpleminded way of checking the power of a telescope they were lucky enough to possess. Such helpful-sounding but actually quite useless directions tend to support Mario Biagioli's contention that Galileo worried about others being able to replicate his telescope too readily and so deprive him of professional credit for its invention. His competitors outside of Italy have already been noted. But a mathematics professor in Rome itself, Antonio Santini, produced a telescope and saw Jupiter's satellites before the end of 1610, as did the Jesuit mathematicians at the Collegio Romano, with help from independent instrument-makers. Galileo's original, hapless intention was to retain the manufacture of the telescope as a kind of trade secret, and his exposed position partly explains the somewhat embattled attitude with which he guarded that secret and defended his recent observations. In turn, what Biagioli calls his "uncooperative stance" may help explain the ferocity of the criticism he sustained from opponents and skeptics.

One of the earliest descriptions of a Galilean telescope is contained in a letter written by Giovanni Battista Della Porta to Marchese (later Prince) Federico Cesi, the founder and principal financial backer of the Academy of Lynxes, an enlightened Roman scientific society critical of the Jesuits. (It derived its name from the lynx's capacity to see in the dark.) Della Porta was the celebrated author of *Magia naturalis*, a volume of technical games and tricks, and *De refractione*, a work on optics (which, while erroneous in itself, dispelled some earlier misconceptions). He told Cesi that he had investigated the "secret of the spyglass" and found it a "*coglionaria*," a vulgar expression for something idiotic; Galileo's invention, he claimed, was "purloined from the ninth book" of his own treatise on refraction, a wholly unfounded assertion. "And I shall describe it," he went on,

so that if you want to make it Your Excellency will at least have fun with it. It is a small tube of silvered tin, one *palmo*, *ad*, long and three inches in diameter, which has in the front a convex glass in the end *a*; there is another canal of the same, four fingers long, which fits into the first one, and in the end *b* it has a concave [glass], which is soldered like the first. If observed with that first one alone, far things are

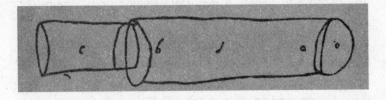

Della Porta's Sketch of Galileo's Telescope (1609)

seen near, but because the vision does not occur along the perpendicular, they appear obscure and indistinct. When the other, concave tube is put in, which gives the contrary effect, things will be seen clear and upright; and it goes in and out, like a trombone, so that it adjusts to the visions of the observers, which all are different.

Aside from the *coglionaria* remark, the description chimes with Galileo's own. And it makes sense. A biconvex objective would have inverted the image: by refracting light rays along a slope, it would send them to the side of the optic axis opposite to the one from which the rays would have struck the retina if traveling through air alone. And this Galileo corrected by reinverting the image with another spherical lens, the eyepiece.

What Galileo (and, increasingly, other people) had by now discovered was that the principal element in a powerful tele-scope was a high-quality objective lens. But it was not an easy item to manufacture. During this period, the technology of lens-grinding necessitated the acquisition of glass blanks, cut from blown globes relatively free of impurities, from which the lenses would be derived; a pedal-operated lathe; forms or grinder heads, convex or concave, which would be used to directly grind the lens; and some kind of abrasive. Since Padua was so close to Venice, a major glassmaking city, one might assume that the apparatus in Galileo's atelier was derived from prototypes used on the Venetian island of Murano, where a glass manufacturer initially supplied Galileo with lenses. But lenses were ground in many cities, and Galileo went at least as far as Florence and perhaps farther in his search for lenses and blanks. No records or diagrams have survived concerning the composition of Galileo's glass, the type of lathe he used, his forms, or his abrasives. Since he was running a small business facing many potential competitors, he kept his own counsel on such matters. It is known that after he mastered lens-grinding he increasingly delegated the task to an artisan named Ippolito Francini.

The composition of Galileo's lenses has been analyzed at the Institute and Museum of the History of Science in Florence, but the results have not yet been published. It has been determined that the greenish or roseate tint of most of these lenses resulted from impurities in the type of sand used in the manufacture of the glass. The lenses were ground on a lathe more or less as one would turn a glass ashtray to hollow out a cavity, except that certain of their properties baffle researchers. Their shape, though always a sector of a sphere, is usually somewhat irregular, especially toward the circumference, suggesting an element of unpredictability in the interface of form and glass—perhaps, for some reason, the artisan's hand was obliged to intervene and rotate the form or the lens. One also wonders what abrasives were used. Glass is a very hard substance, which basically can be ground either with glass itself, that is, with glass particles, or with an even harder abrasive, such as diamond particles in a slurry. Abrasives can be quite expensive and are often under some form of patent or proprietary lock, such as a secret formula, so Galileo had to find one that was both accessible and economical.

Galileo had access to deposits of sand, possibly along the Adriatic coast, containing substantial fossilized sediments of the microorganisms known as forams (foraminifera, or "hole-bearers"). These tiny amoeboid creatures, today much studied though mostly invisible to the naked eye, usually generate a sort of porous shell, known as a "test," and tend to live in marine environments. There are over 275,000 species of forams, with an extraordinary diversity of tests, many quite elaborate and generally composed of calcium carbonate. Sand containing a high density of fossilized forams can be used as an abrasive, and though no one in the Veneto of 1609 yet knew this (Galileo assembled his first microscopes in 1624), certain people were obviously aware that the sand from specific zones could easily scratch glass. Typically, Galileo found out about this and put it to best advantage.

The Museum of the History of Science owns a telescope made by Galileo sometime between 1610 and 1630. It is a wooden tube almost a meter long, bearing cylindrical housings at either end for the two lenses and covered with brown leather, slightly torn where the instrument was mounted on stanchions. The optical system consists of a plano-convex objective (by far the more difficult lens to produce) and a biconcave eyepiece, with the two lenses arranged so that the focal point of the objective can easily be caused to coincide with the back focal point of the eyepiece. Precisely as described by Della Porta, the housing of the eyepiece serves as a draw-tube that can be pulled in and out for focusing. The objective is 3.7 centimeters in diameter and is positioned against a wooden ring with an aperture of 1.6 centimeters, which in turn is held by a smaller cardboard ring with an aperture of 1.5 centimeters—less wide than a thumbnail—and this is finally sealed by a leather-covered ring with a much wider aperture.

Galileo soon discovered this "stopping-down," or reducing of the diameter of the aperture (which corresponds to the diaphragm of a single-lens reflex camera). Initially, he might have sought to compensate for his own vision deficiencies. He might also have wanted to eliminate the flaws around the circumference of the lenses caused by irregular or hand grinding, as well as inherent spherical aberrations—unexplained at the time but noticeable as blurring—caused by the fact that rays striking spherical glass far from the optic axis, the so-called paraxial region, at the center, bend more than those in the narrow bundle concentrated on this region. He must have noticed at once (perhaps from his own squinting) that such stopping-down increases the sharpness of the image, a huge observational advantage. Yet the more he stopped down his lenses in the interests of image crispness, the smaller his apertures became, and these tiny apertures, together with the increasing focal length of his objectives, eventually resulted in the sort of skinny, reed-like telescope that is preserved in the museum in Florence, which can magnify x21 but has a field of view of only 15 arc minutes. All this constituted a marvelous achievement, and some of Galileo's telescopes apparently afforded a magnification of as much as x30. But if one looks through such a telescope (there is an exact replica in the offices of the Institute), one discovers a field of view so small that it cannot encompass the moon—it is not an easy instrument to handle. More importantly, the invention in this form did not permit much technological improvement, because any increase in power would have further reduced its field of view. In fact, Galileo personally made few telescopes after the initial years of discovery, devoting his time to other interests, and eventually the field passed into the hands of much younger men, such as his disciple Evangelista Torricelli; the Neapolitan Francesco Fontana, whom he despised; Eustachio Divini; and Giuseppe Campani. The Galilean telescope was a historical cul-de-sac, for lenses were soon combined in very different ways, mirrors were employed, and, much later, spherical aberration was corrected by the use of aspherical shapes.

Galileo conceived of the perforated plates used for stopping-down as measuring devices, graduated by size, "subtending more or fewer minutes of arc." Indeed the telescope could see a host of new stars, but technical problems in measuring angular distance were posed by the variability of the viewers' eyesight and the minuscule field of view, restrictions not shared by the ordinary quadrant, which could also measure the intervals between stars. Galileo does not seem to have actually used his aperture stops as a measuring instrument.

SOMETIME IN THE late autumn of 1609, Galileo produced a telescope with a power of x20 and undertook, with great excitement, study of the closest of the celestial bodies. The first clear night of such observation initiated a glorious period for Padua and the world, a fitting companion to the period, some three centuries earlier, when Giotto had decorated the nearby Scrovegni Chapel, reinventing the art of painting in the West. Between December 1 and 18, from atop his house, Galileo observed the phases of the moon, also creating seven wash drawings in sepia ink as a record. In his scrutiny of our satellite (or, rather, of the 59 percent of the lunar surface observable before the era of space exploration, the additional 9 percent being revealed by librations, which are very slow oscillations relative to the earth), Galileo was immediately struck by its ruggedness, a plain contradiction of Aristotle, who had insisted on the uniformity and incorruptibility of all celestial bodies. Galileo was also interested in the extremely clear contrast between light and dark surface blotches. The very large dusky patches he called "ancient spots," because they had been seen since the beginning of history by anyone who cared to look—this was an obvious dig at Aristotle; the smaller spots had never before been detected. "The surface of the Moon is not smooth, uniform, and precisely spherical, as a great number of philosophers believe it (and other heavenly bodies) to be," he wrote soon afterward, "but, uneven, rough, and full of cavities and prominences, being not unlike the earth, relieved by chains of mountains and deep valleys." He clearly hoped this observation would put a speedy end to the recurrent theory that the crystalline moon had reflective properties and that its dark patches were only mirror images of the earth's geographic features.

Why mountains? Why valleys? He had studied the terminator, the dividing line between sunlight and shade on the moon's surface. In modern photographs the terminator shows not precisely as a line but as a transition, a zone of half-tone progressively darkening the surface of a sphere, as one might expect. But Galileo, seeing a less magnified moon with a fairly sharp terminator, had noticed a number of bright points just within its shadow side and dark points just within its bright side, and he had sensibly deduced that they were eminences and depressions. On the fourth or fifth day after the new moon, he further noticed that the small bright spots in the dark part increased in size and brightness and, after a few hours, coalesced with the bright part, which had gradually grown larger. On January 7, 1610, he wrote from Padua to Antonio de' Medici in Florence, an illegitimate but influential son of Francesco, the late grand duke. Galileo excitedly announced his improvement of the telescope, disclosing that "one can see most clearly that the moon is not of an equal, smooth, and limpid surface, as she and the other heavenly bodies are believed to be by the great multitude of people, but seen from a closer vantage is actually rugged and unequal, and in sum the moon cannot be concluded in any sane discourse to be other than covered by eminences and cavities, similar to but rather larger than the mountains and valleys distributed around the terrestrial surface."

Galileo's letter to Antonio de' Medici—the first record we have of his observations of the moon—was accompanied by further descriptions, some small sketches, and a quick note concerning some curious "stars" in the neighborhood of Jupiter. All this he would develop further in a pamphlet, written in Latin, still the lingua franca of scientists, whose title, *Sidereus Nuncius*, may be translated either *Starry Messenger* or *Starry Message*, the former having found acceptance. It would appear soon afterward and fascinate the European scientific community. "Now on Earth, before sunrise," he wrote, further explaining his conviction that the moon's surface was craggy, "aren't the peaks of the highest mountains illuminated by the Sun's rays while shadows cover the plain? Doesn't light grow, after a little while, until the middle and larger parts of the same mountains are illuminated, and finally, when the Sun has risen, aren't the illuminations of plains and hills joined together? These differences between prominences and depressions in the Moon, however, seem to exceed the terrestrial roughness greatly." This shows Galileo's likely familiarity with the perspective treatises, often containing illustrations of the lighting of imaginary spiked objects (derived from a type of hat with an interior wicker brace known as a *mazzocchio*) that were proliferating during this period. He also addressed another question. If the moon was as mountainous as he

Sirigatti's mazzocchio

Illustration of a highly stylized *mazzocchio*, a headdress constructed on a wicker arma-
ture. From Lorenzo Sirigatti, *La pratica di prospettiva del cavaliere*, 1596. The understanding of the geometry of the *mazzocchio* came to serve as a demanding exercise for
architects, mathematicians, and students of perspective; its relevance to Galileo's conception of the moon was noted by Samuel Edgerton in *The Heritage of Giotto's Geometry: Art and Science on the Eve of the Scientific Revolution* (Ithaca and London, Cornell
University Press, 1991).

claimed—he did not speak of craters—why do we see its outline as smooth and not serrated? He explained that it was
almost uniformly jagged so that the rows upon rows of serra-
tions would, when observed from afar, obstruct one another
and appear smooth. But Galileo was not content to assert that
the moon had mountains; he wanted to measure their height,
which would strike a fatal blow at the whole Aristotelian
cosmology.

It was an audacious, invigorating idea. For one thing,
nobody in Europe (or outside of the Muslim East) had suc-

ceeded in consistently determining the heights of mountains on the earth. With a theodolite, you could measure the angle formed by your sightline to the peak of the mountain and the plane you were standing on; you knew that an imaginary plumb line dropped from the peak to the base formed a right angle; and you could reasonably estimate the distance in miles from your position to that right angle inside the base of the mountain; all of which gave you enough to calculate trigonometrically the desired result—in theory. The problem was that unless you were looking at a mountain from the sea, such as Mount Tenerife in the Canary Islands, a prized landmark for seamen, it was almost impossible to determine how far the foot of the mountain lay above sea level, and that queered your calculations. In 1644, two of Galileo's disciples, Evangelista Torricelli and Vincenzo Viviani (spurred perhaps by Descartes), had the idea of measuring heights barometrically, since the air pressure in a glass tube containing mercury, if taken to higher elevations, will progressively fall—but this method required climbing the mountain in question. Practical methods for determining mountain heights were not achieved in the West until the nineteenth century, and they entailed lengthy surveying.

It was hard enough, then, to measure the mountains on the earth. Galileo had of course never set foot on the heavenly body whose eminences he was proposing to calculate. But he saw this as an advantage, since some things are easier to measure from afar than from nearby: it's like solving a geometry problem. In this case, he already possessed two constants: the diameter of the moon, computed in Ptolemy's *Almagest*, and the fact that sea level, often so hard to determine when one is inland on earth, simply equaled the moon's spherical surface.

In addition, the terminator of the moon at first or last quarter—the half moon—which describes half the moon's diameter, could be considered, for geometrical purposes, to meet the sun's illuminating ray at a right angle.

Using his telescope, Galileo had noticed that some of the barely lighted points beyond the terminator lay at slightly more than one-twentieth of the moon's diameter inside this area of darkness. He took them as typical mountains. "Accordingly," he wrote, converting ancient Greek units into contemporary Florentine miles and reformulating the problem in plane geometry,

> let CAF be a great circle of the lunar body, E its center, and CF a diameter, which is to the diameter of the earth as two is to seven. Since according to very precise observations the diameter of the earth is seven thousand miles, CF will be two thousand, CE one thousand, and one-twentieth of CF will be one hundred miles. Now let CF be the diameter of the great circle which divides the light part of the moon from the dark part (for because of the very great distance of the sun from the moon, this does not differ appreciably from a great circle), and let A be distant from C by one-twentieth of this. Draw the radius EA, which, when produced, cuts the tangent line GCD (representing the illuminating ray) in the point D. Then the arc CA, or rather the straight line CD, will consist of one hundred units whereof CE contains one thousand, and the sum of the squares of DC and CE will be 1,010,000. This is equal to the square of DE; hence ED will exceed 1,004, and AD will be more than four of those units of which CE contains one thousand. Therefore the altitude of AD on the moon, which represents a summit reaching up to a solar ray GCD

and standing at the distance CD from C, exceeds four
miles. But on earth we have no mountains which reach to a
perpendicular height of even one mile. Hence it is quite
clear that the prominences on the moon are often loftier
than those on the earth.

The Aristotelian postulate of the perfection of heavenly bodies had been confidently proved false at one stroke, though the
actual *height* of the mountains was immaterial.

Certain remarks are in order here. First, Galileo knew by
this time that the moon shows us only one face, positive evidence of the important fact that it spins. (The time it takes
for the moon to revolve around the earth is equal to its
period of rotation about its own axis.) This meant that he
had observed, at best, only half the mountains on the moon
(not counting those concealed by its librations, of which
however he would remain unaware for awhile). He used a
doubtful conversion ratio for the ancient Egyptian *stadia* in
which the earth's diameter was calculated, and took the ratio
of the moon's diameter to the earth's as 2/7 (= 0.2857),
whereas our estimate is 0.2727. Adjusting Galileo's calculations for the modern ratio, but letting his math stand, C. W.
Adams found that a Galilean moon-mountain height would
be at least 5,409 of our miles, or about 8,704,941 meters. The
far side of the moon first became observable with the voyage
of the Soviet spacecraft *Luna 3*, in 1959; if we leave this out
of the picture, however, we may note that the height of the
highest mountain on the normally visible side of the moon,
Mount Huygens, has in fact been determined to be 5,500
meters, whereas the highest mountain south of the Alps that
could be approximately measured then, Mount Aetna,

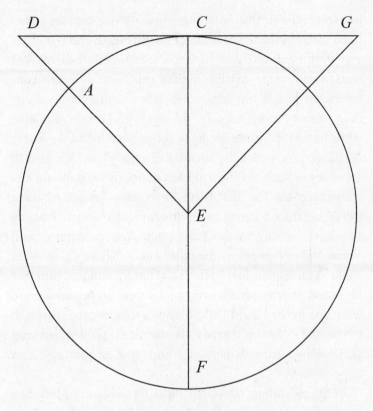

Galileo's Diagram

Galileo's proof of the minimal height of lunar mountains (from *The Starry Messenger*, 1610).

stands at 3,326 meters. (The height of Europe's highest, Mont Blanc, is 4,808 meters.)

BY LATE 1609, Galileo had managed to mount his telescope and train it widely over the skies, and in his long letter of January 7, 1610, to Antonio de' Medici he mentioned something beside the moon. "I have with respect to the stars noted this," he wrote. "First, that with the telescope one can see many fixed stars that otherwise cannot be discerned; and then, this very evening, I have seen Jupiter accompanied by three fixed stars, usually quite invisible in their smallness, and their configuration was in this form: ooo Nor did this form occupy more than a degree, roughly, in longitude." Three weeks later, while he was in Venice seeing to the publication of the *Starry Messenger*, he wrote his friend Belisario Vinta, the Tuscan secretary of state, that he had been struck by a wonderful circumstance: the fact that these three stars—only now there were four, since a fourth had swum into his vision on January 14—were moving in the same plane "as do Mercury and Venus and peradventure the other known planets." He asked Vinta to remember him to the ducal family, and told him that he would shortly send a copy of his new book and a good telescope to the Grand Duke Cosimo, then twenty years old, so that the duke could verify his assertions. Galileo had long yearned to return to Florence, and now he saw a golden opportunity.

In the meantime, on March 13, 1610, the *Starry Messenger* was published in Latin. Its edition of 550 copies fanned out across Europe and made Galileo an international celebrity. In it, he discussed not only the moon and various stars but also the remarkable phenomenon occurring in Jupiter's neighborhood, the one he had mentioned to Vinta. Attention has been drawn to Galileo's exceptional perspicacity in identifying this strange configuration amid the nocturnal excitement of pointing his telescope at hundreds of new stars never before seen, and then in successfully tracking, for weeks, its ever-changing appearance. What he described in the pamphlet, with diagrams, was a constant shifting of these "stars" now to the east or west of Jupiter, but always in a straight line and always "in the line of the zodiac" (that is, the ecliptic, because the twelve constellations of the zodiac, seen from the earth, lie almost totally in the plane of the ecliptic). Since these bodies sometimes either followed or preceded Jupiter by regular distances, and since they accompanied the planet in its retrograde motion, Galileo soon concluded that they were moons. "Here we have a fine and elegant argument," he wrote, "for quieting the doubts of those who, while accepting with tranquil mind the revolutions of the planets about the sun in the Copernican system, are mightily disturbed to have the moon alone revolve about the earth and accompany it in an annual rotation about the sun." He was addressing the many skeptics who believed that if the earth revolved about the sun, it could not retain a moon, but his phrasing was patently prudent, granting a merely suppositional value to the Copernican cosmos. *If* the Polish astronomer's system is valid, Galileo seems to be saying, *then* the moon's revolutions surely track along with the motion of the earth, analogous to those of the Jovian moons.

The telescopic observations made Galileo one of the most famous men in the Western world. In a short period he manufactured many hundreds of telescopes, of which about 6 percent met his specifications—one gathers that the rest were not used—and sent the best by courier to royalty and to the aristocracy: to Duke Cosimo's cousin Marie de Médicis, wife of Henri IV of France, to the duke of Urbino, to the duke of Bavaria, and to the elector of Cologne. Cosimo himself was cleverly persuaded to use the telescope as a diplomatic gift, which turned him into a sort of ducal sales representative. The Jesuit Collegio Romano, which accepted most of Galileo's conclusions (and would later turn against him), warmly received him in Rome. He was inducted into the Academy of Lynxes. Poets sang Galileo's praises in many languages, comparing him to Columbus, to Magellan. The *Starry Messenger* was reprinted in Frankfurt within a few months, and Kepler, the discoverer of the laws of planetary motion, wrote Galileo an enthusiastic letter.

It is never the case that a man, whatever his gifts, can get ahead on his merits alone, without securing any form of sponsorship from on high. This was all the more true in an age without intellectual-property protection, and Galileo knew it. In the past he had sought the support of the Venetian Senate, and now, at age forty-six, as the author of a revolutionary scientific pamphlet, he saw his chance to win a more powerful patron. Jupiter's moons were his to name as he pleased, and he wrote Belisario Vinta on February 13, asking which of two designations, the Cosmican Stars (*Cosmici Sydera*) or the Medicean Stars (*Medicea Sydera*), might best please Duke Cosimo. On February 20 Belisario replied that the latter appellation was—understandably—favored. Flattery got Galileo a very long way. By July of that year, after amiable negotiations with Vinta, he had been hired at an enormous salary as mathematician and philosopher to the court of the grand duchy of Tuscany, outraging the Venetians who had done so much to further his career. It was a fateful move. He would never again live in the Venetian Republic, a state that regarded with a jaundiced eye the Roman Inquisition's jurisdiction in accusations originating outside the Veneto.

Galileo's observations of the moon and Jupiter's satellites did not turn him into a full-fledged Copernican, though they certainly added arrows to his anti-Aristotelian quiver. The turning point came with his observations of Venus. In order to appreciate their significance, however, one must first understand why the orbit of this planet was such a hot issue in 1610. Since time immemorial, observers have watched the sun describe a great eastward circle, the ecliptic, on the celestial sphere during the course of the year. The ecliptic passes visually through the twelve constellations of the zodiac, and it is also, roughly speaking, the plane of the planetary orbits. In ancient times it was noticed that whereas the stars maintain fixed positions, the planets appear to "wander," traveling at times to the east, like the stars, and at other times to the west, the latter motion being called "retrograde." Venus, for exam-

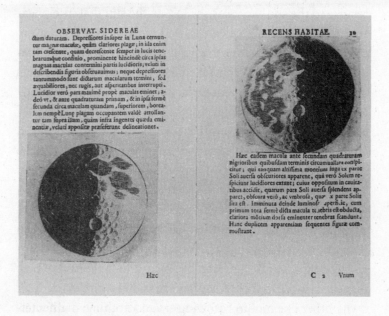

Two pages from Galileo's Sidereus nuncius (The Starry Messenger), *1610.*

These copper engravings were possibly executed in large part by Galileo himself.

ple, makes a sort of loop in the sky, which was accounted for early on by the idea that it described a circle, called an epicycle, which when seen from earth gave it the appearance of going in reverse.

In modern terms, this is explained by the "elongation" of Venus. The angle formed by the line from a planet to earth and the line from the sun to a planet along the ecliptic is called its elongation, and a planet is said to be in "conjunction" when the elongation is 0 degrees, that is, when it passes in front of or behind the sun, forming a straight line. It is in "quadrature" when the elongation is 90 degrees—when a planet forms a right angle with earth and the sun, and in "opposition" when the elongation is 180 degrees, when it passes behind the earth relative to the sun. Venus is so close to the sun that its elongation never exceeds 48 degrees, and since it is an "inferior" planet, closer to the sun than the earth, it can have neither quadrature nor opposition. Of course, the ancient astronomers did not see the solar system as we do now, but everybody noticed that the remarkable thing about Venus, the "morning" and the "evening" star, was that it kept so close to the sun. It could not be at a wide variety of angular distances from the sun, like the outer planets. Where, the ancients wondered, was Venus's orbit?

Ptolemy's *Almagest*, which was still accepted in Galileo's day and which Galileo himself had studied and taught when younger, placed Mercury and Venus as inner or inferior planets circling around the earth between it and the sun. These planets also described epicycles, subsidiary circles of their own, rather like moons revolving around nothing; mathematically, the posited epicycles could be made to square more or less with their retrograde motion as observed from the earth.

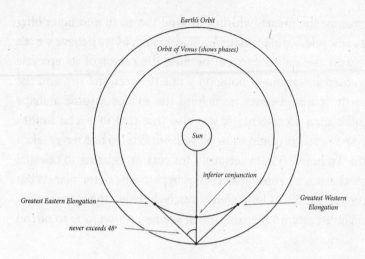

The Elongation of Venus

From this point of view, a principal objection to Copernicus's hypothesis was that Venus, being an inferior planet, should display phases yet did not. (This was also true of Mercury, a trickier case since it is closer to the sun and so harder to observe.) In other words, since in the heliocentric system Venus's orbit lay between the stationary sun and earth's orbit, we should be able to perceive varying phases of reflected sunlight on its globe—crescent, half, gibbous, and so on—rather like the moon's, depending on whether it lay in its eastward or westward elongation. Copernicus had correctly replied that the phases of Venus, though surely displayed, remained invisible to the naked eye.

Unfortunately, this also served as a rationale for the Ptolemaic cosmos, because in that setup Venus would be equally hard to see (though appearing differently from the earth, had we vision powerful enough to perceive it). In the Ptolemaic cosmos, the planet, whirling around the earth and describing its epicycle, would usually be illuminated, as we believe we see it with the naked eye; but because the center of its epicycle tracked unvaryingly along on a line between the sun and the earth, it would never go behind the sun or assume a shape fuller than a crescent. (It was also true that in Tycho Brahe's system, Venus would show phases; but Galileo had never taken the Tychonic system seriously, for reasons relating to celestial mechanics. According to Tycho's hypothesis, all the planets but the earth, including some much heavier than the earth, revolved around the sun, which seemed *prima facie* to offend

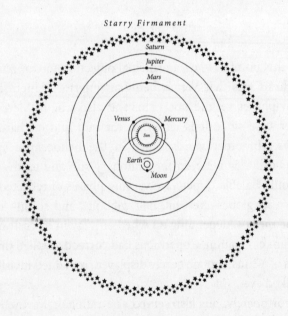

The Tychonic World System

Galileo seems never to have seriously considered Tycho Brahe's cosmology, although it was not at variance with his observations of the phases of Venus.

the uniformity of natural law.) For most of 1610, Galileo was eager to have a look at Venus or Mercury with his telescope, largely to verify whose cosmos was the real one or had the most evidence to support it. To his great frustration, he could not do so, as both planets were usually too close to the sun.

At last, in October of 1610, Galileo began to observe the phases of Venus, and on December 11 he sent an anagram to this effect to Kepler, in Prague—such encryptions were then a common means of protecting an invention or discovery. Kepler failed to decode it, but Galileo's observations were epochal. They gave astronomy its first clear view of planetary motion and, by an intelligent inference, an idea of the revolution of the earth. In essence, Ptolemaic theory, unable to resolve the question of the earth's possible motion, had ended up regarding the planets' perceived motion—planetary motion on the observed celestial sphere—as their real motion. Denying that the earth moved, it failed to see that the celestial sphere, a visual construct, could serve as a useful tool only when astronomers bore in mind that the earth's real motion must be factored into the planets' perceived motion. Though Kepler had worked out the means for computing planetary motion about five years earlier, Galileo was now seeing that motion take place.

If his earlier observations had led Galileo to decisively reject the Aristotelian cosmos, it was his viewing of the sunspots, between 1611 and 1613, that caused him to declare outright his support of the Copernican system. The sunspots themselves had been seen during the realm of Charlemagne, if not before, and three scientists had recently preceded Galileo with similar observations: Johann Fabricius in Wittenberg, Thomas Harriot in England, and Father Christopher Scheiner, a brilliant Jesuit professor of mathematics at the University of Ingolstadt in Bavaria. But Galileo knew nothing of their work, and in any case no one had proposed a valid theory concerning the nature of this phenomenon, though Ludovico Cigoli, the painter-architect, had corresponded with Galileo about the sunspots and had sparked his further interest. Sunspots, as "spots," are actually optical illusions: since the temperature in a sunspot, a complex and impermanent electromagnetic event, ranges from about 4300 kelvin in the umbra, the dark center, to about 5500 kelvin in the penumbra, the lighter surrounding region, the sunspot itself, which is lower in temperature by more than a thousand degrees than the photosphere, appears more or less silhouetted against its background. Fascinated and perplexed by what he could see of this phenomenon, Galileo intensively studied the sunspots from Venice, from Rome, and from a friend's villa in Tuscany during much of 1611.

Galileo's *Letters on the Sunspots*, of 1613, were penned as replies to Mark Welser, a wealthy Augsburg merchant who had sent a paper by Scheiner to Galileo. Scheiner, in publishing his speculations, had had to adopt a pseudonym, "Apelles," for fear of compromising the Jesuits. On becoming acquainted with Scheiner's views on the sunspots, which supposed them to be stars or planets moving at some remove from the sun, Galileo formulated a lengthy reply, which despite its civil tone embittered Scheiner for decades. "I confess to your Excellency that I am not yet sufficiently certain to affirm any positive conclusion about [the sunspots'] nature," he wrote Welser.

The substance of the spots might even be any of a thousand things unknown and unimaginable to us, while the phenomena commonly observed in them—their shape, their opacity, and their movement—may lie partly or wholly outside our general knowledge. . . . Let them be vapors or exhalations then, or clouds, or fumes sent out from the sun's globe or attracted there from other places; I do not decide on this—and they may be any of a thousand other things not perceived by us. . . . I do not perceive the spots to be planets, or fixed stars, or stars of any kind, nor that they move about the sun in circles separated and distant from it. If I may give my opinion to a friend and patron, I shall say that the solar spots are produced and dissolve upon the surface of the sun and are contiguous to it, while the sun rotating upon its axis in about one lunar month, carries them along.

Galileo insisted to Welser, correctly of course, that the darkest sunspot was at least as bright as the brightest part of the moon, and only appeared murky because of simultaneous contrast. He went on to observe that though sunspots moved in a disorderly manner, they traveled as if the sun were carrying them along in a west-to-east rotation; that they were a "tropical" occurrence with respect to latitude; and that as they neared the sun's circumference and passed around out of view, they gave no sign, *pace* Scheiner, of separation or extrusion from the parent body.

Most of these speculations have been verified: the sunspots do emit great light; they are a tropical occurrence, seldom seen at heliographic latitudes greater than ±45 degrees; and they are not detached from the sun. Most importantly, the sunspots as markings could indeed allow astronomers to determine whether the sun rotates, and Galileo was the first to grasp this fact. (In 1860, Richard Christopher Carrington discovered that the sun's rotation varied with respect to latitude, the speed at the poles lagging behind that at the equator by a ratio of about 0.74; the "proper" motion of the slower sunspots was so slow that they traveled from east to west with respect to the sun.)

In *Letters on the Sunspots*, Galileo often freely admits that he does not know or understand one thing or another. It could be, for example, that the sun is not revolving; that only the spots revolve; or that the spots might look different when seen from elsewhere on our globe, though this seemed implausible. In a remarkable passage, he grants that the sun probably revolves and then wonders what causes its motion, and here his earlier research in mechanics connects with his astronomical observations. Aristotle had had no conception of impetus, and thus no conception of motion corresponding to what we may see and measure. He thought that the medium through which objects travel sustains their motion. By contrast, Galileo wrote, "I seem to have observed that physical bodies have physical inclination to some motion," which he then described—lacking the mathematics for an exact characterization—by a series of "psychological" metaphors, themselves of partly Aristotelian origin: inclination, repugnance, indifference, and violence. "Inclination" meant gravity; "repugnance" indicated resistance to being pulled in a direction opposed to the force of gravity; "indifference" referred to a certain tendency of bodies to stay as they are; and "violence" referred to an external force setting something in motion or increasing its motion. Galileo's conception of the sun's motion is necessarily hesitant and ambiguous, and he was wary of flatly stating general principles. But one can perceive here the rough outline of what would become Newton's first law of motion, or principle of inertia, namely, the tendency of bodies to remain at rest or to continue to move with constant velocity unless acted upon by an external force—what Galileo called violence.

In response to another treatise by Apelles, the pseudony-mous Father Scheiner, Galileo again wrote Welser, elucidating his views on the sunspots and on the elongation of Venus. He asserted among other things that the axis of the sun's tilt was perpendicular to the ecliptic, an error he would soon rectify. More importantly, disputing certain of Scheiner's claims about Venus, he claimed in an excess of exasperation that any knowledgeable astronomer who had read Copernicus's *Revolutions* would realize that Venus revolved around the sun and that his understanding would also serve "to verify the rest of the [Copernican] system." Toward the end of the letter, in a discussion of the elusiveness of Saturn, whose form he felt unable to establish with certainty, he suggested in a breathtak-ing phrase that "perhaps this planet also, no less than horned Venus, harmonizes admirably with the great Copernican sys-tem, to the universal revelation of which . . . propitious breezes are now seen to be directed . . . , leaving little fear of clouds or crosswinds." Though technically such confidences to Welser did not imply the public teaching or imparting of heliocen-trism, both men belonged to the Academy of Lynxes, and its president, Federico (now Prince) Cesi, soon published their correspondence. At last Galileo had openly declared for Copernicus. That did not exactly put him in the good graces of the Holy Office.

A number of scholars who had no connection with the Inquisition were infuriated at Galileo's discoveries, or scornful of them. One of these was the Aristotelian Cesare Cremonini, at Padua, who loudly declined to look through the telescope. Another was Giulio Libri, at Pisa, who marshaled textbook arguments to wish away what the instrument plainly showed. A scamp named Martin Horky wrote a letter to Kepler claiming that Galileo's telescope revealed nothing when trained on the heavens. Such examples could be multiplied. The sheer refusal of these men to accept the usefulness of the new research tool, or sometimes even to put their eye to it, reminds us of the classical literary etymology of the Italian word for envy, *invidia*, whose Latin root (-*video*, "I see," plus the privative prefix *in-*, which reverses its meaning) means "not seeing," or "refusal to see," to accord recognition. There was, for awhile, this extraordinary need on the part of many people *not to see*, mostly in the sense of acknowledgment but sometimes in the brute physical sense.

For the Galilean telescope, essentially a tube fitted with glass, the human retina was part of the optical system, and it is possible that certain novice viewers—the *profani*, as they were known—really saw nothing on occasion. Many people, especially scholars, wore spectacles; the sky could be partly obscured; and the diameter of Galileo's eyepiece was, as noted earlier, about 1.5 centimeters. In addition, the eye had to be placed in a central position, which in this instrument was very near the eyepiece, in order that the arriving bundle of light rays could strike the retina intelligibly. This position could be a little hard to find, especially if one was impatient, skeptical, or pigheaded, as some of Galileo's detractors were.

When Galileo reported the motions of the moons of Jupiter, it drove certain people to distraction. They refused to believe that he had seen such a thing. "I must write of a harsh objection leveled at me by all the astrologers and many of the

physicians," a Neapolitan philanthropist wrote to a scholar at
Padua in the spring of 1610. "These people say that if so
many new planets are added to the number of those known,
this will of necessity ruin astrology and demolish most of
medicine, in that the distribution of the houses of the zodiac,
the essential dignity of the signs, the quality of the nature of
the fixed stars, the records of the star-chroniclers, the government of the ages of men, the months of the gestation of the
embryo, oh, a hundred and a thousand things which depend
on the sevenfold order of the planets—all will be destroyed
from their foundation up." The writer wasn't convinced by
these objections: after all, he quaintly noted, since the *luminosity* of Galileo's newly discovered celestial objects was the
same as ever, why would their astral influence increase by one
iota? But one senses the ferocity of the traditionalists' resentment. What the eye could detect with this funny "eyepiece-
reed," or *cannocchiale*, as it was called in Italian, destroyed "a
hundred and a thousand things," reaching into every aspect
of human life. They didn't want to see; the invention had to
be a hoax.

So the question immediately arose as to how Galileo could
secure credibility or authority for his telescopic observations.
Not only did no system of experimental repeatability exist, in
that initially no one else had a telescope of the requisite
power to verify his discoveries, but also he did not want
mathematicians (as opposed to princes) to possess such
instruments. It was a real dilemma. Very soon, of course,
telescopes began to proliferate, in part because Galileo's
invention passed into the hands of other scientists, and in
part because rivals learned how to manufacture them. But in
principle the problem of confirmation remained: six months

after the publication of the *Starry Messenger*, the only witnesses to the existence of Jupiter's moons whom Galileo could produce were the Grand Duke Cosimo de' Medici, Giuliano de' Medici (the Tuscan ambassador to the Hapsburg court at Prague), and himself. One solution, introduced to Galileo by Benedetto Castelli in 1612, and also practiced by a number of others, including Ludovico Cigoli and Domenico Passignano, the painters in Rome, who had obtained a telescope, was to transform the instrument into a primitive camera obscura by projecting the enlarged image onto a piece of paper. Unhappily, this system assumed strong light-collection—the light grasp of a telescope is proportional to the square of its aperture—which was hardly true of Galileo's instrument. So the camera obscura idea, though useful for studying the sunspots, failed to work with the planets and stars. Another possibility was to offer viewing sessions to select notables, but such occasions were so dependent on the weather and on the eyesight and open-mindedness of the notables in question that they acquired a doubtful, séance-like aura. Finally, and perhaps most interestingly, there was the very modern option of the "virtual witness," or pictorial record that could be universally consulted. Single images would have certified nothing and would not have satisfied the Baroque-era demand for a sense of how things moved. But sequential diagrams of phases, rotations, and sunspot movements had every claim to credibility and provided a picture of developments in time. In a way, they were like the frames of a movie. Galileo perceived this almost instantly and offered in the *Starry Messenger* sequential, comparative engravings of the moon, in the first and last quarter. These engravings rendered the terminator as a sharp line to prove his point about lunar geology: one could see the same prominences and depressions lit from opposite directions. He carefully indicated the bright spots falling just inside the area of shadow, the spots he identified as mountains, and it became hard to gainsay his evidence.

(There was an alternative solution to the problem of verification, which never greatly interested Galileo. It was to organize public sessions for independent observation by persons of truly irreproachable integrity, as Kepler did in Prague in September of 1610. Over several nights, using the telescope that Galileo had sent to the elector of Saxony, Kepler and three other viewers of standing confirmed the existence of Jupiter's moons.)

Galileo's wash drawings of the moon, on which the engrav-

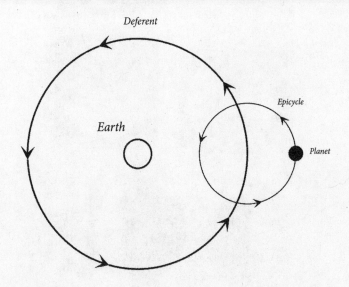

Deferents and Epicycles

Aristotelian astronomers posited deferents and epicycles to explain retrograde planetary motion.

ings were based, excite perennial fascination. Nothing of the
sort had been done before. Botanical and zoological illustration abounded, but the preconditions for astronomical
illustration—being able to see the subject—hadn't previously
been achieved. One imagines how Galileo must have struggled with these little moon pictures, considering his tele-

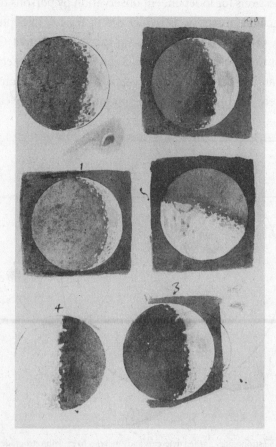

Galileo's Wash Drawings of the Moon (December 1609)

scope's minuscule field of view and the need to constantly refocus his eyes on a sheet of paper; yet this was the birth of an extremely important field: astronomical representation. Until recently, only seven studies were known to exist, all in sepia ink, six on one page and one on another. They were probably first drawn in faint chalk and later rendered in a succession of washes, the chalk traces afterward removed with *mollica*, compressed bread. As might be expected of one who had received a modicum of artistic training, the Tuscan physicist showed in these drawings, which were as delicate as anything one might hope to produce under the circumstances, an ability to capture the proportions of the moon's geological features. But the subsequent copper engravings in the *Starry Messenger*, which Galileo apparently had a large hand in, deliberately altered those proportions. In particular, several scholars have noticed the enlargement of one crater, probably Albategnius, to demonstrate the existence of lunar surface depressions analogous to terrestrial ones (Galileo had the valley of Bohemia in mind). In mid-March of 1611, he wrote to Cosimo de' Medici that he planned a deluxe version of the book with illustrations of the moon in all its phases. But this never came to pass.

Rather recently we have learned more about these drawings. On the basis of a libration of nine degrees vertically measured from Albategnius, Gugliemo Righini established in 1975 that Galileo drew the moon's first and last quarter on December 3 and 18, 1609. Librations are lunar oscillations relative to the earth that result in 18 percent of the moon's face being alternately visible and invisible, 9 percent of this surface being visible at any given time; they are caused partly by the moon's rotation being slightly out of sync with its revolution around the earth and partly by the gravitational pull of other bodies. And another discovery was to come. The March 27, 2007, issue of *Corriere della Sera*, the Milanese daily, revealed that five such drawings, previously unknown to scholars, had been discovered in a first edition of the *Starry Messenger* that had recently come on the New York antiquarian book market. Galileo had drawn them directly onto its pages, so they might represent initial sketches for illustrations to accompany the deluxe edition for Cosimo—or maybe some other edition. These wash drawings have not been made accessible to the public, but the two photographs in the *Corriere* show versions of earlier images and not the beginning, so far as one can see, of a complete illustrated lunation.

Lunar cartography would enjoy a marvelous expansion in the seventeenth century, leaving us documents both informative and beautiful, but Galileo had no part in it. He was busy, his eyesight was poor, and the Galilean telescope's field of view could not be plausibly expanded. He limited himself to savaging other people's verbal and graphic descriptions of the moon, most of which he found contemptible. But one suspects that maps did not interest him much—they had too much distracting detail. Galileo had a deep faith in the epistemological value of images, and like the "ideal heads" of Leonardo, his moon pictures approached the world of Platonic form: they were truth made visible. Enlarging them or adding complications would only have compromised their geometrical integrity.

It is often asserted, in brief references to the subject and on the dust jackets of popular books, that Galileo's observations with the telescope convinced him of the truth of the

Copernican hypothesis and, in fact, proved it. This is, of course, mistaken. Nor did his observations with the telescope demonstrate the truth of the Copernican hypothesis, and he was under no illusion that they did. What they showed, momentously, was that the evidence for the heliocentric theory now greatly outweighed the evidence for the geocentric theory. Galileo was very mindful of the logical axiom known as the law of the excluded middle, which pushes you into one assertion if its contrary is demonstrated to be false, and the Aristotelian worldview (which he rather too easily conflated with the Ptolemaic) began to seem so false to him that he explored every possibility that the only alternative he considered was likely to be true. But if this intuition came closer to confirmation with each passing year, he never found a definitive proof for it. Such a proof could only have been provided by some evidence of the stars altering, even slightly, their apparent positions in the celestial sphere, which would signify that the earth was moving. As Tycho Brahe had asserted, having searched in vain for it with his vast battery of naked-eye instruments at Uraniborg, on the Danish isle of Hven, this evidence did not, or not yet, exist.

The apparent relative displacement of the stars, or stellar parallax, is not the only evidence that the earth moves.* Light from the stars is also affected by the earth's constant motion, arriving at an angle—the "aberration of light"—so that to view a star, a telescope must be tilted in the direction of the earth's revolution. Consequently, when James Bradley first measured

---

* The Doppler effect also proves the earth's motion, though this necessitated the discovery of the wave properties of light and so is not relevant to this period.

this tilt, in 1727–28, a moving earth could be inferred. But the apparent angular shift of the stars against nearby bodies as our planet revolves about the sun had since Aristarchus been regarded as decisive proof for the heliocentric theory, if anyone should be able to obtain evidence of it. Stellar parallax was proved only in 1838, by Friedrich Wilhelm Bessel, a feat that also rendered possible the trigonometric computation of the distance of the closer stars.

One sometimes reads claims to the effect that the Vatican rejected Galileo's heliocentrism because of the inadequacy of his proofs for it, but this is mistaken. Galileo never claimed to offer absolutely conclusive proofs. And the Holy Office, as stated earlier, did not rigorously dispute his science but merely warned him not to champion unorthodox ideas—at least until his condemnation in June of 1633.

GALILEO'S STUDY OF the moon helped validate his long-standing suspicion that the direct observation of nature would undermine the postulates of Aristotelian natural philosophy. But it also ran counter to religious beliefs and folkloric traditions, and one may speculate that offense to those beliefs and traditions stoked the resentment of some of his opponents. As it happened, the moon, identified in Greek antiquity with the goddess Selene and associated with female fertility, had acquired a nearly unassailable status in two branches of Roman Catholic thought. One was the theology of Saint Thomas Aquinas, in which the moon enjoyed an incorruptibility directly derived from Aristotelianism. In his "Three Articles" concerning the fourth day of creation, in Supplement III to the *Summa Theologica*, Aquinas asserts that the heavenly bodies are "living beings" which can "impart life" in their service as agencies of an "intelligent power." Though to question Aquinas was by no means heretical—his opinions were not articles of faith—such probing might be viewed as dangerously rash. Small wonder, then, that in the spring of 1611, when the Jesuits of the Collegio Romano were studying the *Starry Messenger*, whose conclusions they approved on the whole, two things bothered them: the description of Saturn, on purely observational grounds, and the imperfection of the moon. Father Christopher Clavius, the distinguished German mathematician who headed the Jesuit Commission, absolutely insisted on the moon's smoothness, suggesting that any blotching of its surface was caused merely by a tint unequally spread throughout its body.

The other tradition was the rising doctrine of the Virgin's immaculate conception. Though still unorthodox, this enjoyed wide popularity in the early seventeenth century, especially among the Franciscans, whose patron saint, in his *Canticle of the Creatures*, had devoted a verse to "our Sister Moon . . . clear, precious, and beautiful." The fundamental idea that Mary, like Jesus and unlike the rest of us, had been born without original sin was deeply rooted in popular piety by 1609 (though it did not become Roman Catholic dogma until 1854). It stems from the perception of a paradox: that a child, as the product of sexual union, must itself be naturally sinful unless divinely exempted, and that nonetheless Mary, as the mother of God, could hardly have been born of sin. The learned doctors of the Church were divided on the question. Saint Bernard of Clairvaux attacked the idea of Mary's sinless conception as a mere superstition. Saint Thomas Aquinas held that Mary was not exempt from original sin because Christ was the savior of all men, and so Mary had need of His redemption, though he conceded that Mary's soul had been cleansed of her sin some time before her birth. Duns Scotus, a Franciscan, subtly argued that although sinless, she would have contracted original sin had it not been for Christ's intervention, which seemed to reconcile her sinlessness with her need for salvation. For centuries, the doctrine of the immaculate conception was preached by the Franciscans and opposed by the Dominicans, with the Franciscans gradually gaining the upper hand.

That Mary's freedom from sin should be connected in later iconography with the image of the moon belongs to a strain of Catholic thinking, illustrated by much beautiful art, which teaches that shadows or intimations of what will come to pass may be discerned in antecedent revelations. This notion of recurring "types" of human behavior is essential to Christian thought: as the tree of life foreshadows the cross, for instance, so the prelapsarian or initially sinless Eve foreshadows Mary. Though it was hard to dispute the frivolous and probably pagan origin of the Virgin's symbolic association with the moon, the first two verses from Revelation 12 appeared to confirm it: "A great and wondrous sign appeared in the heaven: a woman clothed with the sun, with the moon under her feet and a crown of twelve stars on her head. She was pregnant and cried out in pain." For this reason, the moon, though accorded no cult, became associated with Mary in folklore and honored in later medieval and Renaissance symbolism and devotional painting. Luna was Mary's *impresa*, her heraldic emblem, and to impugn Luna was, in some minds, almost to impugn Mary herself.

Whatever he may have made of such notions (and he

appeared to criticize them *sotto voce* in a letter to Gallanzone Gallanzoni of July 16, 1611), Galileo's favorite work of Renaissance verse, Ludovico Ariosto's *Orlando furioso*, the long heroic poem first published in 1516, which he frequently cited and of which he possessed a heavily annotated copy, features a delightful moon voyage that presages Galileo's own debunking of the crystalline moon. One might even say that rather in the manner of a Christian "type" or prophecy, it strangely foreshadows Galileo's own lunar observations. Ariosto's passage was intended in part as a sort of reprise, in a somewhat burlesque key, of Dante's moon ascent in the *Paradiso* (Canto II, lines 31–36), a passage that Galileo, as a lover of Dante, must have known.

Dante was a serious student of medieval astronomy, and the *Paradiso* was intended to reflect the actual world-picture of Italian savants in the early fourteenth century. Here, as we behold Beatrice and Dante the protagonist ascending into the heavens, he attentively notes the moon's physical nature:

> *It seemed to me that a cloud surrounded us,*
> *brilliant, dense, solid, and unsullied; adamantine,*
> *and as though penetrated by a sunbeam.*
>
> *Into itself the eternal gem received*
> *Us, as water may receive a ray of light*
> *And yet will not divide, but stays united.*

Curiously, the moon's substance, at once solid and cloudlike, is conceived here as entirely enveloping the pair. Without fussing unduly over the five qualities the bard attributes to the moon—

brightness, density, solidity, glossiness, and adamantine hardness—we may note that it reflects standard Thomistic doctrine. The "gem" stands for the moon itself, while the metaphor of light penetrating water without decomposing it, referring to the pair's ecstasy, was also a standard trope for the fertilization of the Virgin's womb.

Now what does the jaunty Ariosto make of the traditional Aristotelian moon? The issue only arises because Orlando, the hero of his poem, has lost his sanity, as many a good knight has done before him. In the thirty-fourth canto of the *Furioso* we hear Saint John the Evangelist tell another paladin, Duke Astolfo, that the only way poor Orlando can ever recover his wits is if Astolfo undertakes a journey to the moon, traditional instigator of "lunacy." So when night falls, the evangelist hitches four horses to a chariot and they climb through a sublunary ring of fire, believed at that time to produce the comets, toward our gleaming satellite.

> *Having crossed that fiery sphere they arrive*
> *at the realm of the moon, which looks like a steel plate,*
> *entirely spotless, and about the same size, I've*
> *been told, as the earth—that would include our great*
> *oceans which add to our globe. After the drive,*
> *which has not taken long, I would estimate,*
> *Astolfo expresses his astonishment and surprise*
> *that the moon, which looks small from earth, is of such size.*
>
> *The other unexpected revelation is that*
> *the earth is hard to discern, emitting no light*
> *by which to perceive it. From the distance at*
> *which Astolfo is standing the earth is quite*

indistinct, its pleasant habitat
not projecting far out into the night.
With difficulty he can get some hint
of where the oceans are—but he has to squint.

On the moon there are rivers and lakes and hills and dales,
Like those we have but different. And also towns
With houses and public buildings, but on such scales
As we are not accustomed to. He frowns
In amazement and concentration, for earth pales
In comparison. There are also woods and downs
Where nymphs and fawns are hunting fierce moon beasts
And celebrating afterwards with feasts.

Duke Astolfo does not pause to explore
every feature of the moon. He is
there to transact business which is more
pressing. The Apostle is aware of this
and leads him downward to a valley floor
where all that we have lost or has gone amiss
through Time or Fortune or our own grievous fault
is collected and stored as if in a huge vault.

I do not speak only of realms and gold
That Fortune's unstable wheel gives or takes back.
But also those things beyond what she can hold
Or give—Fame, for example, which the attack
Of Time can devour before it has grown old.
Up there as well are countless prayers our slack
belief has offered up, and vows that were broken
very nearly as soon as they were spoken.

Ah, and the tears and sighs of lovers, and all the hours
That gamblers lose and ignorant men waste,
The plans that we make that are well within our powers
But require perseverance if not haste
Before they fade away. And books of ours
That we intended to study rather than taste.
Those can often be heavier losses than
Material things in the life of any man.

Among other lost or mislaid items that the apostle and Astolfo find on the moon are women's charms, amorous vows, flopped love affairs, threats, royal crowns, gifts and flattery lavished upon princes, disastrous plots, villains' schemes, charity postponed until the benefactor's death, and—finally— a large flask, full of a "thin liquid, apt to evaporate," containing Orlando's wit. So Orlando will recover his sanity. Leaving aside Ariosto's other astronomical details, such as the fact that the earth reflects so little light, what Galileo must have remembered in 1609 was the moon's initial appearance in these stanzas as a uniform "steel plate" and its subsequent revelation to be a body possessing "hills and dales, Like those we have but different." One can imagine him smiling wickedly at these verses as he calculated the minimum height of a lunar mountain. Yet, in a development at once delicious and dismaying, on reading the *Starry Messenger* Kepler wrote Galileo a letter describing in loving anthropological detail the likely attributes of a lunar civilization somewhat similar to Ariosto's: "They have, as it were, a sort of underground city. They make their homes in numerous caves." The level-headed Galileo demurred: He had seen no clouds upon the moon's face. Without water, how could there be life?

IN THE MEANTIME, the Catholic Church was proceeding apace with another system of representing the heavens, one that was not scientific but imaginative and propagandistic. The Church had invested an enormous capital of intelligence in astronomy, partly for calendric and doctrinal reasons and partly because certain churchmen, often Jesuits, happened to have the mental gifts and education to pursue the subject. It was largely squandered with the injunction against Copernicanism in 1616 and with the Collegio Romano's espousal of the Tychonic, or geoheliocentric, system. Yet as the construction of domed churches proliferated and architects and designers acquired the same mathematics that astronomers used—that is, Euclidean geometry and trigonometry, including spherical trigonometry—it became possible to enlist a specialized branch of perspective to create the illusion of celestial space on the inner surface of a hemisphere or similar shape. Thus, the tradition of the *cupola dipinta*, or painted dome, came about, together with all the correlative ambiguities of the *sottinsù*, the view from below looking upward.

Both domes and planetariums had been erected since antiquity, and the parallel between the church dome and the celestial hemisphere—the apparent firmament in which we perceive the stars, with the observer's zenith shifting as he moves—was perfectly obvious. The circumference of the base of the dome was the metaphorical equivalent of the astronomer's celestial equator. Within the *spazio divino*, or consecrated space of the cupola, it fell to the commissioning prelate, in tandem with the painter and his team, to agree on some manner of representing heaven, virtually always as an ecstatic apparition. By the early 1500s, small frescoed cupolas of this type were already appearing in

northern Italy, but the first great painted dome was executed by
Pordenone in 1519 for the Cathedral of Treviso. Between 1520
and 1530 Correggio in Parma created *Vision of Saint John at Pat-
mos* for San Giovanni Evangelista and *Assumption of the Virgin*
for the Duomo. The latter in particular, with its vision of heaven
opening to a vast angelic flutter, set the tone of future ceiling
frescos, while the expanding girth of the domes themselves
reflected a growing awareness of the size of the universe.

Many early painted ceilings showed concentric images of
heaven, whose processional circularity, reminiscent of a vast
carousel, bore an eerie and ironic resemblance to diagrams of
the Copernican solar system, especially when centered on a
light-transmitting lantern. (A lantern is the glazed cylindrical
or polygonal structure that usually sits atop a large dome.) I
have found no evidence that the general shift during Galileo's
lifetime to dome-painting characterized by a degree of asymmetrical turbulence, observable notably in Giovanni Lanfranco's *Assumption of the Virgin* in Sant'Andrea della Valle, in
Rome, of 1625–27, betokened a conscious attempt to depart
from this "Copernican" concentricity (and actually the concentric scheme continued in use). The painted dome, with its
gravity-free angels, saints, and cherubs, lasted about three and a
half centuries as a living art form, defying Newton's law of gravity, and its morphology does not follow a linear progression.
However, it passed through critical moments of renewal, and a
hinge figure in its evolution was Galileo's friend Ludovico
Cigoli, an architect and painter of sacred history whose work,
like that of Federico Barocci and the Carraccis, anticipated
major features of the Baroque style.

In 1610, Cigoli was fifty-one years old, five years older than
Galileo. He would die in 1613. Partly because of Galileo's

Cigoli's Dome

Ludovico Cardi Cigoli's fresco of the Madonna Immacolata in the cupola of the
Pauline Chapel of the Basilica of Santa Maria Maggiore, in Rome (completed 1612).

influence, he had become convinced of the truth of the
Copernican system. Cigoli felt that those who persisted in
upholding the doctrine of the crystalline moon not only erred
scientifically but revealed their impious envy, for *invidia*
implied not only the refusal to see what genius had discovered
but also what God had created. On March 18, 1610, Cigoli,
then living in Rome, wrote Galileo that

upon the arrival here of Sig.r Ferdinando Martelli, I am
fairly obliged to salute you with this missive; and receiving a
letter written to me from Florence by Sig.r Amadori, to
rejoice with you who have raised your telescope to such perfection that it has been able to perceive and observe marvelous things in the heavens; and I read that you have given
a lecture of some sort concerning all this [a reference to the
Starry Messenger] and have been to Venice to have it
printed. . . . Finding myself as it happened with Sig.r Cardinal del Monte [Caravaggio's principal patron, a supporter of
other artists, and an optics enthusiast] and turning to this
topic, I read him Amadori's letter, and he at once ordered
one of his envoys in Venice to procure the said telescope,
and as soon as the book should be printed to send it to him.

Between 1610 and 1612, Cigoli, corresponding with Galileo
and experimenting with perspective projections upon curved
surfaces, created the *Immacolata*, a superb fresco in the cupola
of the Pauline Chapel of Santa Maria Maggiore, the major site of
Marian devotion in Rome. During this period he and his colleague Domenico Passignano, took time off to peer through a
telescope they too had obtained, sometimes using the great
church as an observatory from which to view the monuments
of Rome and, eventually, celestial bodies, and notifying Galileo
of their sightings. (Passignano saw the sunspots before Galileo
did, using a colored filter to protect his eyes and describing
them as "lakes or caverns.") Cigoli's *cupola dipinta* is not merely
concentric in structure but dynamically rotatory in that its
shapes suggest a whirling motion, perhaps inspired by his read-
ings in astronomy. The one static element is the immaculate
Virgin, who stands upon a crescent moon clearly pitted and

Cigoli's Madonna

Line drawing of the figure of the Madonna in Cigoli's fresco in the Pauline Chapel of Santa Maria Maggiore. She stands on her emblem, a crescent moon, but it is pocked with crater-like cavities, a violation of the Aristotelian doctrine of the perfection of celestial bodies.

scarred, just as Galileo had drawn it. So Cigoli's moon, floating in a chapel in anti-Copernican Rome, is defiantly un-crystalline and un-Aristotelian, and geologically akin to Galileo's, though it does not depict any part of the moon's actual geography.

Throughout the late sixteenth and seventeenth centuries, dome after dome was erected and painted to inspire faith and awe, and to hold out the promise of salvation. But if the Church

hoped that these lofty extravaganzas would offer devotional tran-
quillity, that hope was frustrated by weird perspective problems.
Had the imperatives of religious decorum and, increasingly, a
certain theatrical seduction not been paramount, the problems
could have been readily solved. But the designers soon recognized that given the demands of visual legibility, they were very
hard to solve, for within these great temples one had to consider
mobile observers in relation to huge hemispherical or polygonal
ceilings, and in their proposed solutions a curious perceptual relativism crept in that was cousin to some of Galileo's discoveries.
Galileo argued that our perception of moving bodies was subject
to a kind of relativity, because we ourselves conducted our observations from a moving body, the earth; now a somewhat analogous relativity cropped up with respect to our perception of very
large hemispheres.

We are not comparing two mathematical models here, but
juxtaposing a mathematical model and an expressive art form;
it is, however, the optical rather than the aesthetic properties of
the painted dome that concern us. And we see at once that an
aesthetic problem, of a socio-religious nature, generated a host
of difficulties for the dome-painter. Suppose your aim as a
dome-painter is to create a ceiling that works as an illusionistic
extension of the worshipper's actual space—you want to co-opt
him, to include him in a moment of prearranged ecstasy. If so,
the entire perceived design of the room, both real and fictive,
will have to function as a continuum, no matter where on the
floor the observer stands. And this is indeed possible in a very
small room, such as Andrea Mantegna's Stanza degli Sposi in
Mantua, with its little trompe l'oeil cupola. As soon as you're
dealing with a big church dome, however, two problems arise.
First, the total area of the wraparound hemisphere is larger

than the arc subtended by the angle of human vision, so that wherever the observer happens to be, even at the center, some of the painted architecture will appear anamorphic, that is, distorted, and certain of the painted figures will almost certainly appear upside down. Second, an accurate perspectival projection of figures on the dome will leave them too small to have any psychological impact, and too spatially compressed to be comfortably legible. Your teensy saints and angels will diminish to dwarfishness as they ascend toward your vanishing point, which is heaven's zenith, and they will lack the foreshortening that includes them in the observer's world. You, the dome-painter, know that such effects will undermine both the religious decorum and the sensuous enchantment that the Counter-Reformation has come to embrace.

What could be done? Well, the theorists of the Baroque devoted a lot of thought to these problems, culminating in the invention of *quadratura*, a branch of perspective devoted to projecting geometry onto vaults and domes (perspective being a system of drafting rules derived mathematically from geometrical optics). But all such designing, whether aimed at depicting a concentric or an asymmetrical heaven, faced the same persistent problem of figures that were too small and too spatially compressed. In response, some painters intuitively concluded that the composition on the dome should reflect not the real observer's viewpoint, which sent the painted heaven too far away, but a virtual viewpoint much closer to the ceiling, which allowed for much larger and more legible figures. As such a viewpoint ruptured the illusionistic continuity of the dome's space with the observer's space, however, and left it convincing only if seen from one part of the floor, the idea arose of using balustrades, grilles, or other devices to confine

the observer to this optimal position. One of the major expo-
nents of *quadratura*, Andrea Pozzo, inserted a marble disc in
the pavement of a Jesuit church, the Sant'Ignazio di Loyola in
Rome, whose ceiling he had painted, and indeed from this
viewpoint the illusion is overwhelming; but as one moves
away, the soaring trompe l'oeil architecture creates a dizzying
effect and threatens to cave in. Pozzo was well aware of this

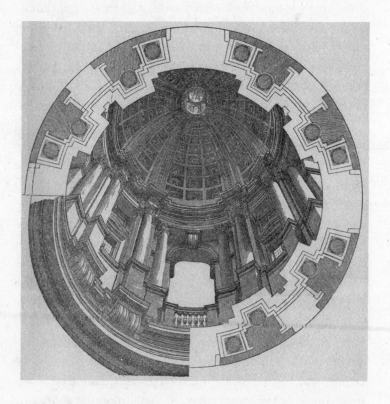

*Andrea Pozzo's Design for the Cupola over the Crossing
of Sant'Ignazio di Loyola, in Rome (painted ca. 1685).*

Pozzo stipulated that the cupola was to be seen only from the vantage point implicitly
indicated by this view, the vanishing point being equated with God's glory.

illusion, and at the end of his life wrote an influential treatise on *quadratura* in which he pleaded for a unifocal, off-center, stationary viewpoint. Harking back to what the art historian Samuel Edgerton has called "the traditional medieval belief that God spreads his grace through the universe according to the laws of geometric optics," he exhorted painters to view the perspective vanishing point, or infinity, as an expression of God's glory. Pozzo's marble disc symbolized all too clearly and unhappily the doctrinaire standpoint of the Collegio Romano vis-à-vis mathematicians like Galileo and Kepler. Just as the

*Computerized, Anamorphic Image of Pozzo's Cupola When Seen
from a Peripheral Viewpoint Not Authorized by Him*

Actually, the cupola looks even stranger than this when seen from beneath its circumference, but today it has darkened so much that it cannot be usefully photographed. Galileo despised the anamorphic effect, and in fact the binding of viewers to a fixed point under the dome came to serve as a metaphor for the Jesuit refusal to accept Copernican reasoning.

Jesuits insisted that the earth stood still, so they also pinned observers to a single point in the church of Sant'Ignazio: move, and everything collapses. It was the perfect visual symbol of a gigantic intellectual self-deception.

Meanwhile, some painters opted for the alternative strategy: diffuse, dissociated imagery designed to be seen successively from a moving axis. Mattia Preti seems to have decided that a multifocal image could be best enjoyed by a mobile observer whose itinerary imaginatively coincides with an array of virtual viewpoints. This approach also acquired many adherents: in Giovanni Battista Piazzetta's *Glory of Saint Dominic*, a late example in Santi Giovanni e Paolo in Venice, one can make out no fewer than five scattered vignettes on the ceiling, all bathed in the same golden soup, yet each with its private vanishing point.

There is, in this tacit admission of the need, under any vast frescoed ceiling, to circulate in order to align oneself with multiple focal axes, a Baroque parallel with Galilean relativism, though of course it concerns optics, not celestial mechanics. Yet if Galileo's astronomical contribution consisted in turning the Copernican theory into an actual model, capable of verification, the Baroque church-dome was not a rival model: if it had been, it would have resembled a gigantic armillary sphere in stone. It was, rather, a metaphor, something of a pious distraction, and as the century progressed, the decorative skills it enlisted were unabashedly derived from those used for the stage. It must be granted that many people no longer find frescoed domes very compelling: of all the excesses of the Baroque style, these neck-bending heavenly carnivals are among the least popular, and even devout believers may find them too extravagant to touch the soul, which is what they were intended to do. But they are marvels of engineering, design, and virtuoso painting, and they represent the Church's liveliest response—if only an implicit one—to the threat of Copernicus and Galileo.

GALILEO'S DISCOVERIES AND essays had not failed to produce an effect. On February 24, 1616, alarmed by the spread of writings sympathetic to the idea of the earth's motion, the Inquisition issued a decree condemning Copernicanism as "foolish and absurd in philosophy, and formally heretical" because it contradicted Holy Scripture. On March 3 of the same year, the Congregation of the Index—the branch of the Holy Office responsible for censorship—declared that two such works, Copernicus's own *Revolutions* and Father Diego de Zuñiga's *Commentary on Job*, were to be "suspended until corrected." A third, the *Letter on the Pythagorean and Copernican Opinion of the Earth's Motion*, by the Carmelite friar Paolo Antonio Foscarini, was "prohibited and condemned." Galileo was not mentioned, presumably out of respect for the Tuscan ducal family and the prestige he enjoyed with the Jesuits. In case he should fail to get the message, however, he had been summoned a little earlier, on February 26, to the residence of Cardinal Robert Bellarmine, the head of the Inquisition and the greatest Roman Catholic theologian of the period, to receive a special injunction. The minutes of this meeting reveal that Bellarmine warned him to completely abandon his Copernicanism, and "henceforth not to hold, teach, or defend it in any way whatever, either orally or in writing; otherwise the Holy Office would start proceedings against him. The same Galileo acquiesced in this injunction and promised to obey." In the meantime, during much of 1615 and 1616, the Inquisition

investigated Galileo on suspicion of heresy but found little in
the way of solid evidence, and the case was dismissed; since
the charges were revived in 1633, however, we will examine
them shortly.

It is a chastening task to read through the documents pertaining to Galileo's relations with the Church in search of
some longed-for, blessed text in which Copernicanism is substantively discussed. Revealing, though hardly substantive, is
the implicit verbal duel between Bellarmine and Galileo that
took place when Galileo was being investigated. Saint Robert
Bellarmine, or San Roberto Bellarmino, as the Italians call
him (he was canonized in 1930), is generally regarded as the
most incisive champion of the Counter-Reformation. A tiny,
gentle, exquisitely well-mannered man, he had written what
the Vatican regarded as the definitive refutation of the Protestant heresy, the multivolume *Disputations on the Controversies
of the Christian Faith* (1581–93), whose careful argumentation
had earned even its opponents' respect. He had lectured at
Louvain, practically on the front lines of the wars of religion,
and had spoken out eloquently against English King James I's
anti-papal Oath of Allegiance. As a Jesuit cardinal he had
served on the tribunal that condemned Giordano Bruno to
death—the free-thinking philosopher was burned at the stake
in the Campo dei Fiori in Rome in 1600—though Bellarmine
later confessed to being plagued by remorse over Bruno's
refusal to repent and receive last rites. Bellarmine had reached
his early seventies by 1616 and was nominally bound by the
Jesuits' "rules of study," which closely followed Saint Thomas
Aquinas; yet his own theology hewed much closer to the literal meaning of Scripture than had that of Aquinas, a grave
misfortune for Galileo.

Robert Bellarmine, in a letter of April 12 warning Friar Antonio Foscarini to abjure his heliocentrism, explicitly cited Galileo along with him. In the third numbered point of this elaborate missive, Bellarmine stated,

> I say that whenever a true demonstration would be produced that the sun stands at the center of the world and the earth in the 3rd heaven, and that the sun does not rotate around the earth, but the earth around the sun, then at that time it would be necessary to proceed with great caution in interpreting the Scriptures which seem to be contrary, and it would be better to say that we do not understand them than to say that what has been demonstrated is false. But I do not believe that there is such a demonstration, for it has not been shown to me. To show that the sun in the center and the earth in the heavens can be made to match our astronomical calculations—this is not the same as to prove that in reality the sun is at the center and the earth in the heavens. The first demonstration, I believe, can be given, but I have the greatest doubts about the second. And in the case of doubt one should not abandon the Sacred Scriptures, as interpreted by the Holy Fathers.

To this argument Galileo silently responded with a ferocious memorandum, which today goes by the title "Considerations on the Copernican Opinion," in which he explained that neither Copernicus nor he regarded the heliocentric setup as suppositional at all, but in all probability as factually true. Luckily for Galileo, this piece never got farther than Prince Cesi, who apparently advised him not to send it to the head of the Inquisition.

Bellarmine's words merit close analysis. Elsewhere Bellarmine had pointed out that to deny the spoken is to deny the speaker, which in the case of the Scriptures is God; so far so good, one follows his reasoning. To this divine authority he adds that of the Holy Fathers, on the grounds that the Church as an institution is inerrant, which is already going rather far; but again one follows his thought—after all, the cardinal represents the embattled Counter-Reformation. What happens next, however, is fascinating. Bellarmine makes a perfectly correct distinction between mathematical or suppositional astronomy and physical astronomy, as though he were quite familiar with such matters, and declares himself willing to accept a purely suppositional Copernicanism. But what if somebody like Galileo were to insist that the sun, as a physical body, really occupies the center of our world? That is where the trouble starts. If a sound proof were offered him, Bellarmine says, he would have to reinterpret Scripture; but can a sound proof actually *be* offered? Of this he has the greatest doubt, and when in doubt, he relies on Scripture, as interpreted by the Fathers. These are circular sophisms, as the Italian philosopher Guido Morpurgo-Tagliabue has written, which "under the pen of Cardinal Bellarmine reveal that to command it is not necessary to reason: in fact, one loses the habit." Bellarmine's objection to Galileo is the first formulation of the fundamentalist objection to science. It fails to grasp both the scientific concept of a hypothesis and the fact that nature can be known only through mathematics.

If we continue our search, we discover that a few years later, in 1619, Galileo became embroiled in a controversy with Father Orazio Grassi, a Jesuit and later the architect of Sant'Ignazio, over the origin of the comets, which Grassi held to be a

fatal flaw in the Copernican system. This dispute is now practically impossible to follow, but Grassi was ill-informed scientifically, did not represent the Vatican, and did not accuse
Galileo of any theological deviance.

Then there is the Ingoli case. Buoyed by the election to the
papacy in 1623 of his old friend Maffeo Barberini, poet and
humanist, Galileo decided in the following year to reply to a
brief letter in essay form sent to him in 1616 by a priest
named Francesco Ingoli; this piece was titled "De Situ et Qui-
ete Terrae" ("On the Position and Stability of the Earth").
Ingoli had recently been placed in charge of the Vatican's
institute for the propagation of the faith, the Propaganda
Fide, but neither in 1616 nor in 1624 did his opinions on
Copernicus formally represent those of the Catholic Church;
indeed, there is every reason to suspect that the new pope
himself had encouraged Galileo to rebut Ingoli. Though uniformly mistaken, Ingoli's arguments against Copernicus were
thoughtful and serious, and sidestepping all theological
issues, Galileo politely refuted every single astronomical position in Ingoli's eight-year-old letter in a long, cogent polemic
called the "Reply to Ingoli." Among other errors, Ingoli had
misunderstood the computation of parallax; had taken it into
his head that the center of a circle was the farthest point
inside it from any point on its circumference, which Galileo
disproved with a geometrical demonstration; and believed
that the earth must be at the center of the universe because
the "denser and heavier of simple bodies [occupy] lower
places"—a formulation that, with the word "lower," assumes
the very issue to be proved. "Where do you get the idea,"
Galileo asked Ingoli, "that the terrestrial globe is so heavy? . . .
For me (and, I believe, for nature) heaviness is that innate

inclination by which a body resists being removed from its natural place, and by which it spontaneously returns there when it has been removed by force." Ingoli has several other, more sophisticated objections, which Galileo would later take up in the *Dialogue*, and though Ingoli speaks for himself alone, and not for the Vatican, his essay-letter is the closest the Church came, before 1633, to a substantive engagement with Galilean astronomy. One wishes for more; there ought to be more; but amazingly enough, that is all there is.

Galileo's ongoing dispute with Orazio Grassi led in 1624 to his writing of *The Assayer*, his popular and beautifully written essay on the modern scientific outlook. Upon publication the title page was redesigned by the Academy of Lynxes to bear a dedication to the new pope, who liked the book exceedingly; but he could not have failed to notice Galileo's deceptive attack on Copernicus's idea of the earth's "third motion," that is, a motion in addition to rotation and revolution, whereby the earth maintains its axial stability and whose existence Galileo contested. Slyly, Galileo said nothing about the earth's other two motions, "smuggling in" (to use Stillman Drake's phrase) his tacit concurrence.

The Tuscan physicist was now sixty. The Roman political climate seemed favorable to his opinions, and he began to contemplate the writing of a major defense of the heliocentric world picture. He was encouraged in this ambition by Maffeo—Pope Urban VIII—who looked forward to it as a brilliant exposition of an interesting computational system that had nonetheless failed to dislodge the approved Aristotelian cosmology. Galileo was given to understand that his book would prove to the civilized world that Catholic Italy was fully abreast of the latest developments in mathematics but had declined to adopt heliocentrism because of its scientific and theological shortcomings. Perhaps Galileo never fully understood the pope's expectations, or perhaps he forgot his own job description somewhere along the line. In a way, this was understandable: after all, the Inquisition's "corrections" to Copernicus's *On the Revolutions*, applied in 1620, amounted to only thirteen passages that failed to specify the system as suppositional (and many extant copies from that period, including the one in the very library of the Collegio Romano, show no emendations). It was reported, moreover, that the pope had laughed out loud at dinner at Ciampoli's reading of *The Assayer*, which made open sport of the Jesuits. So Galileo slowly began work on the huge, self-incriminating manuscript that would become the *Dialogue Concerning the Two Chief World Systems*; after long, intermittent efforts, the book appeared for sale in Rome in May of 1632.

By the summer of that year, the Inquisition was once again investigating Galileo, this time on grounds that turned into the two implicit counts of an unwritten indictment. The first was that he had hoodwinked the Vatican into issuing him a license to publish the *Dialogue*, a point we will come to later on. The second was that the new book championed Copernicanism, which, as we have seen, was a formal offense since 1616. Considering, then, that Galileo did indeed write the book to suggest the likely truth of the heliocentric thesis, and that this was the charge leveled against him, the modern reader might expect it to adopt a straightforward polemical line. One might assume, for example, that it would offer a complete exposition of the heliocentric cosmos, and then rebut to objections; or that it would take the opposite tack, and demolish the geocentric cosmos, supplanting it at once with that of Copernicus; or that it would juxtapose the two, and reveal which suited the facts better deductively and empirically—one can imagine a number of obvious outlines for the book. But Galileo followed none of them, because they would not have interested him. Nor would they have made much sense in the 1620s.

Galileo chose the dialogue form, spun out in the Italian vernacular rather than Latin, the international language of science, for three important reasons. First, in showcasing his literary skill, he hoped to reach a large audience whom he might captivate and entertain. Second, he could place his scientific arguments in the mouths of fictitious characters, and so perhaps dodge ultimate responsibility for them. Finally, by adopting a modified form of the Platonic dialogue, he could deploy a variant of the Socratic method, a form of conversational teaching that enlists the antagonist's own power of reason to convince him of the truth of an idea. Galileo apparently believed, like Plato, that the mathematical shapes of the world and those conceived by man's mind are corresponding reflections of each other. Human intelligence, properly honored, consists in a search for the simplest and most beautiful geometrical forms that will fit and explain the empirical facts, and as Socrates maintained in certain of Plato's dialogues, we human beings possess these forms somewhere in the recesses of our memory. True philosophy consists in remembering what is, at bottom, inborn knowledge. In this sense, the argument with those who believe in the earth-centered universe is really an attempt to induce them to remember what they intuitively know but have forgotten. The most poignant moment in the *Dialogue* occurs when Salviati, the champion of Copernicanism, draws out of his opponent Simplicio, the exponent of Aristotelianism, the long-submerged, almost dreamlike awareness that the earth circles the sun.

As it happens, however, the dramatic back-and-forth of the conversation is punctuated by another of the author's aims—to eliminate certain objections he has encountered—because along with the teasing Socratic method, Galileo resorts to a fair number of pugilistic jabs. Indeed, all the quasi-journalistic sparring may at first bamboozle the contemporary reader. So it is that we find Galileo's characters addressing the question why, as the world turns, flocks of birds are not left behind in the air; the problem makes us smile but had to be disposed of promptly, as it probably troubled many perfectly sensible readers. The upshot of these overlapping complications is a large, entertaining, but very difficult volume, which it would not be opportune to explicate here, least of all in the form in which it was written. Certain of its arguments were fallacious, as a number of readers, including Fermat, already noted in the 1630s. Yet it also disclosed, by fits and starts, its author's philosophy of science.

It is as an odd truth that one sometimes discovers revelatory aspects of Galileo's methodology by noticing what he does *not* do. It seems logical, for instance, that he would use telescopic observations as the cornerstone of his argument. He doesn't, and for good reason. The most important observations, such as the sightings of the phases of Venus, are in fact reconcilable with the Tychonic system, and if presented sequentially, as they would have to be, they would run the risk of sequential dismissal. The *Dialogue* is an attempt to give its readers a whole new world picture, which entails changing their habits of mind, so Galileo declines to offer them a set of empirical facts, however startling. He wants first and foremost to convince them that doing mathematics is not a petty intellectual chore, analogous at best to the measuring and computing a carpenter must perform when building a house or repairing a boat, which is what a lot of philosophers thought at the time. Repeatedly the Aristotelian interlocutor, Simplicio, is scolded for his lack of mathematical agility in approaching various problems of physics, which require geometrical expression. Of course the attempt to conceptualize problems involving mass and velocity in the terms of Euclidean geometry now strikes us as archaic, if undeniably ingenious, but that doesn't diminish the force of Galileo's insight. Galileo sees experience as a mental snare unless situated within a larger frame of reference: the horizon, after all, bisects the heavens, an experience that ought to place us at the center of things but doesn't. He knows that practically every advance in astronomy in the past hundred years has violated common sense. Certain of his most important laws, such as the law of fall, cannot in fact be verified. "For the Aristotelians," as William R. Shea has written, "this is an insurmountable barrier. For Galileo, it merely proves that the frontiers of science are not coterminous with the frontiers of experience." Only an imaginary experiment makes the law of fall possible, yet only that law can account for what happens to actual falling objects. "One can, indeed one must, go beyond sense experience, but this presupposes a philosophical conviction that the real is described by the ideal, and the physical by the mathematical." Of course we are at the threshold of a philosophical issue here—how we can legitimately claim to have proved anything—that can hardly be encompassed by this book. I might add that scholars have also discussed whether the sources of Galileo's "conviction" lay primarily in Plato; in some sort of quasi-mathematical aesthetics; in his

Christian faith; or—and this is most likely—in some delicate blend of all three.

In the early fourteenth century, Dante achieved a fusion of astronomy and theology in the *Divine Comedy*. His system consisted of nested spheres, within which the Inferno could be elegantly described as a series of conic sections, which Galileo as a very young man had explained in two lectures to the Florentine Academy. Galileo, who loved Dante, did not read him as science—we will return to this point later—but Dante's moral vision had such a firm hold on most literate people's imaginations that they couldn't quite shake it as a descriptive cosmology. And this cosmology, which owed much to that of Aristotle and Saint Thomas Aquinas, pictured the world as a sort of colossal tool for humanity's instruction and guidance. The zodiac, for instance, was relatively close to men and women and governed their fate at God's behest, like a sort of divine keyboard. Did the Copernicans situate the stars at a vast distance from Saturn, the outermost planet then known? Galileo has his Aristotelian character Simplicio recoil at the thought of the intervening void. "Now when we see this beautiful order among the planets," he asks, ". . . to what end would there then be interposed between the highest of their orbits (namely, Saturn's), and the stellar sphere, a vast space without anything in it, superfluous and vain? For the use and convenience of whom?" At such a remove, the stars would have no earthly purpose. Though the earth is a place of corruption, it remains at the center of everything, and the universe exists for the salvation of men's souls.

We have all had the experience of being unable to solve a spatial problem, in any realm of activity from map-folding to chess to auto mechanics, until we have abandoned a blocked

approach in favor of some sudden, fresh insight. Galileo had
observed that the ingrained habit of picturing the earth as stationary made it difficult even for receptive listeners to follow
Copernican reasoning. Etched in their brains was what Shea
has termed a false "graphic representation" of the heavens, in
which the observer continued to visualize his viewpoint as
motionless. Philosophically, this representation came from
Aristotle and Ptolemy; psychologically, it came from Dante.
According to this picture, for instance, planets like Mars and
Venus appeared at certain points in the year to reverse their
course against the background of the stars, which the Ptolemaic system explained by assigning them epicycles like little
gears: this could explain their apparent rearward motion,
though what caused them to revolve around empty space
remained unclear. In place of this jumble, Galileo's Copernican spokesman, Salviati, urges the reader to think in terms of
a different graphic representation. If he would only forget
about looking at the celestial sphere from terra firma for a
moment, and situate himself at a point on the moving earth,
he would see why Mars, which revolves around the sun more
slowly than the earth, and Venus more rapidly, would seem at
times to travel backward.

One by one, Galileo's protagonist demolishes the typical
Aristotelian arguments of the day. An objection to the rotation of the earth is that if it were true, a projectile fired to the
west would travel farther than one fired to the east. Salviati
disagrees and suggests a thought experiment. Mount a crossbow on a carriage and shoot an arrow first in the direction of
its motion and then against it: the arrow will partake of this
motion or be hindered by it. Let us say that the arrow would
travel 400 meters if shot in the direction of the carriage but

only 200 if shot in the opposite direction. If we grant, for the purposes of argument, that the carriage travels 100 meters during the time of the arrow's flight, the arrow will in either case land 300 meters from the carriage's new position. The same is true of any projectile fired on the surface of a planet in motion—it lands an equal distance from where it was shot.

Galileo's aim in the *Dialogue*, as he put it, was to purge the world of the Ptolemaic system with a medicine distilled of the Copernican. But he never found the panacea, the decisive proof, he sought. For a moment, a device in the church of San Petronio in Bologna seemed to provide what he needed. This conjoined a peephole in an exterior wall with a line of marble inlaid in the interior pavement running along the earth's north-south meridian line; it showed that the sun's image, passing through the hole and striking the meridian line, had slightly altered its position at certain moments in the past few years. The measuring "instrument," the church, was unreliable in that it could have shifted (as a result of subsidence, for example), but Galileo worked out a geometrical way of detecting such a possibility and compensating for it. What interested him as a Copernican was the hint of a decrease in the obliquity of the ecliptic, which would indicate some sort of movement in the polar axis relative to the background of the stars: such a decrease would inevitably push the sun's image a tiny bit farther south along the meridian line. But the measurements were inconclusive, and Galileo had not developed a notion of the "wobble" of the earth's axis (which does in fact take place). So he alluded to the meridian line only in passing, noting that he had high hopes that the data collected at San Petronio might prove more reliable in the years to come.

In the end, Galileo came to rely heavily on a solar theory of the tides. There is evidence that he had been thinking a lot about fluid dynamics ever since some long discussions he had with his friends Giovanfrancesco Sagredo and Fra Paolo Sarpi in Venice, a city filled with water, in the mid-1590s. Now, in the *Dialogue*, he worked out a cohesive tidal theory. In essence, it had to do with the earth's "two motions," about its own axis and around the sun. To conceive of the action of the tides, he devised an extraordinarily brilliant mental model, which, however, lacked the mathematical notation (at the very least, analytic geometry and algebra, though in the long run calculus would have been necessary) to attack this problem. It was as though he were using a meta-language, Italian, to refer to a mathematical language that had not yet been invented. In any case, fellow scientists soon noted that his model was insufficiently analogous to the way the seas actually shift on the terrestrial globe.

From the start, Galileo had refused to consider the idea that the moon's differential gravitational action on the seas might cause the tides (with, as is indeed the case, some lesser pull from the sun). If this strikes us as eccentric, we must remember, as Shea has observed, that "the idea of gravitation or attraction was embedded in a philosophy which made much of sympathies and antipathies, of occult forces and mysterious affinities," and that Galileo "had worked himself out of this universe of discourse and was in open revolt against it." We have seen that for many Italian Catholics during this time the moon had a kind of magical aura, and in throwing out the magic Galileo threw out a more plausible explanation of the tides.

It was not, however, the truth or untruth of Galileo's tidal theory, or of any other theory in the *Dialogue*, that caused the Holy Office to summon him to Rome in the autumn of 1632. The Inquisition and the pope were not concerned with verifiability but with obedience. An edict against Copernicanism had been issued in 1616. Had Galileo defied it? Had he championed the banned cosmological hypothesis? And if so, could he be induced to mend his ways and abjure it?

3

The Trial; or, Not Seeing

At the time of the summons to Rome, Galileo was still philosopher and mathematician to the Grand Duke Ferdinand II of Tuscany. We may wonder therefore why the duke never raised a finger to save one of his highest-paid and most prestigious courtiers. Treachery? Pusillanimity? No: Ferdinand, then twenty-two, was neither perfidious nor cowardly; rather, the offhand betrayal of Galileo fell into an already established pattern of Tuscan subservience to the Vatican. Discussions of Galileo's trial seldom raise the question of his extradition, which is accepted as an inevitability to be passed over in silence. In fact, it was such an inevitability, but how it came to be so requires some explanation.

The grand duchy of Tuscany, though constrained like any small principality to pursue a cautious foreign policy, possessed the formal attributes of sovereignty. Florence was then suffering a political decline, but Ferdinand theoretically was an absolute ruler, and in the absence of a concordat between Tuscany and the Papal States, the surrender of one of Europe's leading scientists to a tribunal of the Holy Office at first seems self-destructive on his part, an act of suicidal lèse-majesté. After all, it wasn't as if the ecclesiastical realm, whose territory extended over a portion of central Italy, enjoyed great respect in the peninsula—far from it. Sneered at by the Venetian Republic and threatened by Spain, which ruled Naples and the south, it often found its agencies thwarted or mocked. Venice imposed severe limits on the jurisdiction of the Roman Inquisition within its boundaries, assigning many cases of blasphemy and witchcraft to its own secular courts, insisting that all inquisitorial judges be citizens of the Serenissima, and tending to reject denunciations and testimony proceeding from outside the Veneto. The Roman Inquisition was also seriously hampered at Naples, whose people had rioted in 1547 against the introduction of the Spanish Inquisition, and secular rulers elsewhere meddled in its activities. If Galileo had remained at Padua, on the Venetian mainland, it is doubtful that he would ever have fallen into the hands of the Roman Inquisition, especially in that he stood accused not of heresy but of "rashness," a lesser offense.

At first, Galileo and his allies tried to secure a change of venue, to the Inquisition's office at Florence, where he might defend himself orally and in writing. He was unwell; a plague was rife in some Tuscan provinces; ever short of funds, he feared the expense of detainment in Rome; and the vague possibility of a so-called rigorous examination, or torture by the *corda*, a variant of the rack, theoretically hung over him should he be deemed to be withholding evidence. That putative secrets or intentions can be extorted through the application of pain is an ancient and universal idea. Though torture was practiced far more readily by civil courts than by the

Roman Inquisition, it was far from unknown in this religious context.

Fully aware of the peril Galileo was running, Ferdinand did nothing to protect his most famous subject. Yet the grand duchy had not always been Rome's obliging bedfellow. In the standoff between Venice and the Papal States, for instance, Tuscany long tried to remain discreetly neutral. In 1575, as the Counter-Reformation was being implemented, the arrival of three bishops dispatched to Tuscany to oversee the application of the decisions of the Council of Trent met with resistance, especially from the local clergy, and the reigning Medici grand duke dissolved a society of pro-Inquisition vigilantes who were hounding the townsfolk of Siena. Then there was the sorry dispute between the Vatican and the Este family, lords of Ferrara. A splendid walled city in the Po Valley, known for its frescoed palaces and lavish state banquets, Ferrara had distressed the Papacy by appearing likely, during the 1550s, to fall under the magnetism of a Calvinist princess, Renata di Francia. The Estensi were never forgiven for this little lapse. They had patronized great painters and poets and brilliantly redesigned the entire town, but in 1598, when their succession fell into doubt—it was the classic issue of bastardy—Pope Clement VIII elbowed them out and claimed their domain. An orgy of papal looting followed. Attempts on the part of Ferdinand I of Tuscany to rescue the Estensi, with whom he was allied by marriage, were foiled by Spain and France.

During the Counter-Reformation, no principality, and indeed few significant public figures, could ignore the wishes of Rome, Spain, and France and survive. Besides, the Papacy, if courted, was a source of political legitimacy; the Vatican's aid had been enlisted in transforming Tuscany into a grand

duchy in 1569. Some major popes had been Medicis themselves, or otherwise Florentine, or closely allied with great
Florentine houses. And so a habit of deference grew in the
early seventeenth century, a sort of coziness which the
eighteenth-century Tuscan historian J. R. Galluzzi would
describe as "feeble acquiescence" to the papal court and a
"pernicious tolerance" of its scheming. The Tuscan property
held in mortmain by the Church grew by leaps and bounds,
and the Inquisition snooped about unopposed. By the time
Ferdinand II reached majority and assumed power in 1628, he
was bound by paralyzing conventions: hearing in late 1630,
for instance, that the inquisitor of Siena had been arrested for
arming his retainers in defiance of the law, the poor duke simply let him go, at the behest of the inquisitor of Florence. The
following year, Ferdinand, although he had married the duke
of Urbino's daughter, did nothing as Urban VIII (in a move
analogous to the ouster of the Este family from Ferrara) formally annexed Urbino to the Papal States. Ferdinand was in
no position to oppose Galileo's extradition.

It fell chiefly to the lot of three other men to help the
embattled genius. One was Andrea Cioli, the grand duke's secretary of state, who tried to monitor information from his
perch in Florence. More crucial were Francesco Niccolini, the
Tuscan ambassador to the Holy See, and the pope's correspondence secretary—and Galileo's former student—Monsignor
Giovanni Ciampoli.

Niccolini was the most influential of the three. Now nearing fifty, he had served as a page in the Florentine court and
studied as a novice, abandoning his ecclesiastical ambitions to
marry Caterina Riccardi—the Riccardis were a great Florentine family—in 1618. He became Tuscan ambassador to the

Vatican in 1621, fondly hosting Galileo during his triumphal visit to Rome in 1623–24. Niccolini united in his person three prominent characteristics. He was a Tuscan, totally loyal to the duke and his homeland, and wary of the papal court; he was a diplomat, who could tell the possible from the impossible; and he was a friend, which Galileo, a difficult man in the best of circumstances, could not say of everyone.

Giovanni Ciampoli was a trickier case. It is hard to judge someone's character on the basis of scanty documents, but the records seem to show Ciampoli as a man whose ambition gradually detached itself from his initial field of endeavor to become a free-floating hunger for power. He met Galileo in 1608 and studied mathematics under him in Padua before taking holy orders, in Rome, in 1614. With his considerable talent for poetry, he attached himself to the circle of Maffeo Barberini, who, as we know, was an accomplished poet himself. Without question there is a resemblance between Ciampoli's verse and Maffeo's, especially in their common emphasis on the fleeting nature of earthly delights. Any lover who follows the fugitive forms of joy, Maffeo wrote in a famous couplet, would find such fruit bitter in his hand; Ciampoli, though he deemed the world a "theater of marvels," likewise wrote that he was "weary in my suffering"—meaning not personal suffering but the harshness of confinement in a sensory world. Politically, Ciampoli came to occupy a pivotal point between the Vatican and the Jesuits, whom he knew well but quietly loathed. After Maffeo's papal election, he managed to become not only the pope's confidential secretary but also an important member of the Academy of the Lynxes. The Lynxes were at daggers drawn with the Society of Jesus, and Ciampoli (who, among other diplomatic achievements, managed to improve relations between the pope and Cardinal Richelieu) worked tirelessly to promote Galileo's standing with Urban VIII and to reduce the influence of the Jesuits in the Vatican.

It was Ciampoli who obtained the imprimatur for Galileo's *Assayer*, and who later urged the Tuscan physicist to heed the pope's encouragement to write the more comprehensive work that became the *Dialogue*. His letters to Galileo often have a saccharine, fawning tone, as if he were keen to bask in the light of the great man's friendship. In late 1623, he drafted a letter from the pope to Grand Duke Cosimo that referred to Galileo as "my beloved son." Yet while toadying to the pope, Ciampoli came to despise him, and his encouragement of Galileo despite the pope's anxieties about Copernicanism betrayed the irresponsibility of the self-regarding meddler, the manipulator whose social perceptions are defective and dangerous. Fortunately for later generations, however, Ciampoli, like Cioli and Niccolini, wrote informative letters during the Galileo affair, and through them we may follow the scientist's fate.

From the outset, Ambassador Niccolini and Secretary Cioli were convinced that Rome's concern about Galileo was the consequence of a "calumny of . . . envious and malicious persecutors," especially considering Galileo's undisputed Catholicism. It was perhaps a matter of no serious concern. By September 5, 1632, however, we find Niccolini growing very worried. He writes Cioli that he has spoken at length with Urban VIII, and

I began to think, as you so rightly say, that the world is coming apart. While His Holiness was talking about these distressing matters before the assembled Holy Office, he

flew into a rage and suddenly told me that "Our Galileo has
been too eager to enter where he does not belong, in matters more grave and dangerous than should be raised at this
moment." I countered that Galileo had not published without his own ministers' approval, and that I myself, to that
end, had obtained and sent the book's preface here. He
replied in the same apoplectic rage, saying that Galileo and
Ciampoli had circumvented him, and that Ciampoli in particular had impudently told him that Signor Galileo wished
to do everything that his Holiness commanded and that
everything was fine, and that this was all he knew, without
his having actually seen or read the work.

And so the furious pope went on to regret ever having anything to do with such people as his own correspondence
secretary and the Vatican censor, who had strangely let
such a book slip through their hands. None of this pre-
vented the ambassador from humbly and bravely beseech-
ing the pope to let Galileo defend himself, which he felt was
only fair; to which the pope replied, "In such affairs of the
Holy Office, one only censures, and then calls upon the
accused to recant."

But Niccolini, as one Tuscan to another, did not give up.
"Doesn't it seem to your Holiness," he persisted, "that Galileo
ought to know in advance all the questions and criticisms and
censures being leveled at his work, and just what is troubling
the Holy Office?" The pope was appalled. "I tell you," he shot
back in a rage, "that the Holy Office does not do such things,
and not in such a way, nor does it ever give anyone such things
in advance—it just isn't done! And besides, he knows very
well where the difficulties lie, if that's what he wants to know,

because we ourselves have discussed them with him and he has heard them from our own mouth!"

Dispirited by this encounter, Niccolini advised Cioli that Maffeo Barberini had appeared so ill-disposed toward Galileo that they had best turn to somebody else, perhaps Maffeo's nephew, Cardinal Francesco Barberini. If one tried to oppose the pope frankly, Niccolini said, "he gets his back up and shows no respect for anyone. The likeliest course would be to win him over with time, to skillfully work on him, without causing a stir, and also to go through his ministers, according to the nature of the business at hand."

The diplomat had tried to nudge the autocrat away from his inclination to rely on the sort of procedure that we now call (without reference to the Inquisition) "inquisitorial," in which magistrates compile evidence instead of hearing the argumentation of two adversarial teams. He had suggested instead the application of what were, in fact, the norms of canon law, which would be recognized 150 years later, in a different form, as basic civil rights: the prisoner's right to know the charges against him, and his right to defend himself. In the following days, however, Niccolini learned that such rights were not likely to be honored, because the pope was particularly incensed at the *Dialogue*'s intrusion into matters of faith and was disposed to turn the matter over to the theologians of the Inquisition. Niccolini's most forceful plea, to the effect that Galileo after all was mathematician to the grand duke and universally respected as such, fell on deaf ears. The pope replied that despite his friendship with Galileo he had warned him off these topics sixteen years ago, and now Galileo had wandered into a dreadful thicket and had only himself to blame. When Niccolini quietly suggested that Galileo could hardly be seen to be straying from the basic dogmas of the Church, and that his case had arisen only because everyone in this world has his enemies and invidious detractors, Maffeo snarled, "*Basta, basta!*" quite as if he suspected that the diplomat, who tended to veil his phrases, might be insinuating something about Maffeo himself.

Maffeo Barberini was certainly angry at Galileo, and Niccolini and Cioli and Ciampoli probably wondered how much personal animus there was in his anger. It was true that in 1615, when still a cardinal, he had told Galileo through Ciampoli and another friend, Piero Dini, a former archbishop now living in Rome, and very likely even in person, that though he admired his scientific work, he should stay clear of theology and "speak cautiously, like a professor of mathematics." Yet Maffeo and Galileo had indeed corresponded and been on the most cordial terms. When in the summer of 1623 Maffeo's nephew Francesco had received a baccalaureate and been accepted into the Academy of Lynxes, of which Galileo was a member, the scientist had sent him a letter of congratulation. Maffeo had written back, noting "my great esteem for you" and "your affection toward me and my House, . . . and my ready disposition to serve you always."

Now, in the autumn of 1632, the pope felt strongly that Galileo had taken up Copernicanism in a manner explicitly forbidden and, abetted by Ciampoli, had hoodwinked the Vatican censor. Something else irked him too, something pertaining to the *Dialogue*'s contents. This work, which the inquisitors would later closely examine, does not, as mentioned earlier, attempt to prove the truth of the Copernican system but instead plays certain arguments for it against the geocentric, which turns out to be more vulnerable to polemics. The dialectical schema allowed Galileo to claim, as he did in his eleventh-hour preface, that the whole thing was a kind of intellectual exercise, a demonstration that the Catholic philosophers of Italy knew all the best arguments in favor of Copernicanism and had in the end rejected them, but only because of "reasons that are supplied by piety, religion, the knowledge of Divine Omnipotence, and a consciousness of the limitations of the human mind." Unhappily, this lame preface was printed in a typeface differing from that of the rest of the book, which revealed it as an afterthought, according in substance with nothing in the body of the dialogue but its equally lame conclusion. For in the last few paragraphs, out of the blue, Simplicio, the character who all along has been gamely championing the geocentric cosmology, states in abbreviated form Maffeo Barberini's own theological view on such matters, which Galileo had probably heard directly from the pope around 1624. This view held that God's power to create the world is limited in one way only, namely, that it must obey the law of contradiction—the logical axiom stating that no proposition can be simultaneously true and untrue—and that otherwise God is omnipotent. Consequently, because He can create any number of worlds, we cannot hold with absolute certainty that the earth revolves about the sun. Thus the entire *Dialogue* is bracketed by two passages that attempt, however feebly and bizarrely, to unsay its own arguments.

Much has been written about this dialogue, but almost all of it conflates or confuses three distinct issues. First, it has sometimes been claimed that Simplicio is a literary caricature of the pope, an extreme contention that does not follow from the sense of the text and would assume that Galileo was a self-destructive madman. It has also been claimed that the pope's vanity was stung by seeing his theological opinions placed in the mouth of the callow and student-like Simplicio, and this is certainly true, and also perfectly understandable. Finally, Maffeo Barberini's theological idea is generally taken to be silly or sophistic, but in fact it is not, nor would it have seemed so to Galileo. The inquiry into how we know what we know, and how our knowledge may relate to God's omnipotence, had a long pedigree in scholastic philosophy and was in the process of being taken up by Descartes, among others. The trouble is that it is a metaphysical concern, logically anterior to scientific investigation, and to bring it to bear on celestial mechanics represents what we would now call a category error. Prevailed upon by others, Galileo at the eleventh hour used it to get himself off the hook, with the consequence that a category error became an insult to the bishop of Rome.

So it was that Maffeo Barberini, both doctrinally dismayed and personally offended, turned the Galileo affair over to the Holy Office. On October 3, Galileo learned that his request for a change of venue had been rejected. His presence was urgently requested in Rome.

ALL WAS NOT lost, however. An extraordinary letter from Galileo was delivered on October 13, in Rome to Cardinal Francesco Barberini, Maffeo's nephew, whom Galileo thought of with reason as a friend. The missive, of the favor-currying sort, complained bitterly of the hatred to which the great man had been subjected by his rivals, who had been "thrown into shadow," as he characteristically put it, by the "splendor" of his scientific writings. Galileo informed the cardinal that his enemies had somehow prevailed upon the Inquisition to suppress the *Dialogue* and, more alarming still, that the dread office expected him to appear that very month before its tribunal. "This torment causes me to rue all the time I have spent in my studies . . .," he lamented. "And in making me regret having shown some of my work to the public, it may induce me to suppress and burn whatever remains in my hands, thus satisfying the yearnings of my foes, whom my ideas so greatly annoy." For all his high self-regard, persecution was turning him against himself.

Francesco, a friendly, meditative man with a pointed mustache and a little pointed beard, was entirely unlike his uncle, from whose political decisions he programmatically dissociated himself. Temperamentally an aesthete, he had become a great francophile while serving as papal legate to Avignon in the 1620s, and he owned several pictures by the young Nicolas Poussin. In 1625, he had bought what soon became Palazzo Barberini, on the Quirinal, and had assumed responsibility for its decoration. With the help of Cassiano dal Pozzo, he turned its library into one of the finest in Rome.

Galileo nursed reasonable hopes that this sensitive soul would take up his cause, or at least deftly mediate between himself and the pope. His letter does not appear to have furthered these expectations, but the cardinal gave proof of forbearance, and perhaps a dram of sympathy, in refraining (so far as we know) from disclosing the missive's contents to the Inquisition, for it went on to make some very strange and intemperate claims. Galileo, in lieu of going to Rome, offered to put all his ideas in writing, which would redound to the glory of the One True Church—which was fine in itself, but he didn't stop there. He went on to say that he had been convinced of the value of his offer "when I heard a saintly and admirable pronouncement issue from the mouth of a person most eminent in doctrine and worthy of veneration for the sanctity of his life, setting forth in no more than ten words (strung together with keen-witted loveliness) as much wisdom as may be found in all the discourses of the sacred doctors . . . I will for now withhold its admirable author." Was the mathematician claiming to have visitations from the other world? Or was he referring to the pope himself? The cardinal might well have wondered. Farther along in the same wildly rambling, deeply melancholy letter, Galileo insisted that "if what I write should fail to mitigate whatever bill of indictment is brought against me, when objections are raised I shall not fail to respond as God dictates to me." *As God dictates*? For much of the previous century the phrase "dictated by the mouth of God" had been used by churchmen to refer to the Bible. So if Galileo in an unguarded moment was claiming to receive revelation in the manner of Moses or Paul, Cardinal Barberini kindly let it pass. He did, however, forward the letter to his uncle.

Niccolini must have had word of this, for soon afterward he wrote Galileo a shrewd note, warning him not to defend himself but to "portray yourself as the Cardinals of the Inquisition see you; otherwise you will have the greatest difficulty in defending your case, like many others before you. Nor, Christianly speaking," he went on, in his diplomatic doublespeak, "can you claim anything other than what they, as a supreme inerrant tribunal, may wish." And having assumed a role very like a modern public defender angling for a plea bargain, he went off to haggle with the pope.

"I tried to awaken in him some compassion for poor Signor Galileo," he told Cioli,

> now so old and loved and venerated by me, all on the assumption that his Holiness had seen the letter [Galileo] had written to his nephew the cardinal. His Holiness replied that he had seen the letter but could not waive the requirement that Galileo come to Rome. I then suggested that his Holiness ran the risk of forfeiting his case whether here or there, since Galileo's age, together with his extreme discomfort and misery, might cause him to expire on the way. He replied that he could come slowly, in a litter, with all due comforts, because he wanted to examine him personally, and may God forgive him the error of getting involved in such a tangle after his Holiness himself, when still a cardinal, had strongly cautioned him to stay out of it.

After which the pope had again proceeded to attack his *bêtes noires*, Ciampoli and Niccolò Riccardi, the Vatican censor (who in fact was Niccolini's wife's cousin), calling them flunkies who brazenly thwarted their master. So Niccolini left empty-handed. And in the end Galileo received nothing from the nephew in scarlet but a promise to shorten the quarantine at the border crossing, which happened to be under his purview.

Between them, Niccolini and Cioli now foresaw two dangerous eventualities. One was that Galileo, with his bumptious self-belief and blindness to his own provocative tendencies, might further arouse pontifical ire or stir up the Roman wasps' nest. The other, more imminent, was that he might die on the way. He was sixty-nine years old—they thought or claimed he was seventy—afflicted by terrible arthritic pains in his midriff and lower extremities and by a bad heart, vertigo, incipient blindness, migraines, a hernia, and chronic depression. Worst of all, the *peste nera*, the Black Plague, had been rife in northern Italy since 1629 and had been creeping into Tuscany as well.

The pestilence, manifesting as buboes and *carboni negri* (which one takes to be coal-black carbuncles or patches of some sort), had incubated in conditions of war. The issue of the Mantuan succession and the pope's incompetent interference in the crisis had lured the Hapsburg Empire into Lombardy, dragging able-bodied men out of the countryside and causing a shortage of grain. Famine was compounded by drought. As fields were left fallow, Mantua and other cities were sacked by the imperial troops and many of their inhabitants slaughtered or raped. Hordes of filthy vagabonds began to besiege the intact towns and villages, and disease soon spread among the encampments of refugees. When the pestilence itself inevitably arrived, probably brought by the undisciplined German soldiers, the authorities began to isolate the sufferers, yet still the Black Plague raged throughout Lombardy, harvesting hundreds of thousands of lives. It crept south toward the papal principality, where a quarantine of forty days was imposed on those intending to cross into papal territory. At times in various parts of Tuscany, women and children, and sometimes men as well, were forbidden to leave home for long periods, and on some days all but those delivering foodstuffs were barred at the gates to Florence. The plague never ravaged Tuscany as it did the north, but only in May of 1633, when the Madonna of Impruneta, an icon of the

Virgin supposedly painted by Saint Luke, was borne in solemn procession from Impruneta into Florence, did the contagion die away.* Though on December 17 Galileo garnered three doctors' attestations to the effect that the journey to Rome posed a danger to his survival, the Inquisition insisted on his presence, and it was decided that he would travel by closed litter and undergo a quarantine reduced (thanks to Cardinal Barberini) to eighteen days. His exposure to the plague was thereby diminished, but given his fragile state, the slow journey would entail a health risk of its own, as would the long delay at the border crossing of Ponte Centino. The forced journey from Florence to Rome was, for a sick old man, a form of torture in itself, or at the least, a display of what lawyers today might call depraved indifference: there was some question even then as to whether the juridical procedure—extradition, detainment, interrogation—would not be more painful than any sentence eventually handed down, a plain violation of canon law. Was Maffeo Barberini vindictively chastising Galileo in order to salve his wounded pride? So Niccolini and Cioli might have wondered, though they couldn't do so in their letters. In the Tuscan ambassador's correspondence, however, the word that stands out whenever he recalls his conversations with the pope is *aggirato*: over and over, the pope complained that he has been *aggirato*—gone around, circumvented. He was infuriated, outraged, galled that so many people—Galileo; the Vatican censor; Ciampoli, his own scamp of a letter scribe who has made common cause with a hostile, pro-Spanish camarilla in Rome; and heaven knows

* This plague, one of the worst ever to affect Italy, is described in a famous passage in Chapter 39 of Alessandro Manzoni's *The Betrothed*.

who else—have "gone around" him. In Niccolini's portrait of the pope we have a picture of power without charisma, without any graceful, natural capacity to command. As a ruthless Tuscan politician, Maffeo readily yields to spite—one recalls Niccolini's remark that he has "no respect for anyone"—but his dismay also appertains to his role. As Urban VIII, leader of the Catholic Church and chief administrator of the Counter-Reform, he legitimately worries that doctrinal chaos is erupting on his watch, and he has no clear view of what is happening. Though his motives are mixed, this is something that no manager can tolerate, and only the Holy Office, under his eye, can bring it back under control.

At this point, one thing was certain: Ciampoli, who had been conspiring all the while on Galileo's behalf, had it coming. And late in November, another old friend of Galileo's, Benedetto Castelli, now mathematician to the pope, told him that Ciampoli had been disgraced. Castelli was a former student of Galileo's at the University of Padua, a Benedictine priest, and an eminent scientist in his own right. He wrote, a little naively, that Ciampoli, having "astonished all Rome with his frank spirit and shrewd conduct," had left the city for an insignificant governorship at Montalto della Marca. Though all Ciampoli's friends were upset at this development, the former correspondence secretary was behaving like a Stoic, not only unbowed but quite as though he had never been involved in a struggle at all. "Fully self-possessed, he's more light-hearted than ever, applies himself to his studies, and, best of all, actually shows irreproachable reverence toward the Church fathers, submitting quietly to the will of God." Visiting the inveterate schemer in his study, Castelli found him in a state of perfect composure, troubled only by Galileo's ordeal.

As for Galileo himself, Castelli said that since he had never failed Mother Church in any way, "your malignant persecutors wish for nothing better than your arrival in Rome, so they can raise cries from the ignorant mob and call you rebellious, insubordinate. . . . For that reason alone, you should vigorously resolve to hold your own against the weakness of age and the inclement season, and gird yourself for the journey. . . . And recommending yourself to God, sir, may you come joyfully, too, because I do believe you will surmount every obstacle."

Galileo left Florence on January 20, 1633, traveling in a litter provided by Grand Duke Ferdinand, an expense he was later asked to cover out of his own pocket. After submitting to the quarantine, he arrived in Rome on February 13, lodging at the Villa Medici with Francesco Niccolini and his wife, Caterina.

READERS WHO HAVE visited Rome will recall that the Villa Medici stands on a hill called the Pincio, which commands a magnificent view of the city. Built for a Medici cardinal about sixty years before Galileo's arrival in 1633 to face trial, the villa had soon afterward become the Tuscan embassy. In this edifice, Galileo was nobly but uncomfortably lodged, since its stairways proved trying for an elderly man with severe arthritis. Held under house arrest, he was not allowed to leave the grounds of the embassy unescorted until early March, when he received permission to stroll in the fragrant garden beside the church of Santa Trinità dei Monti. Apparently, he never entered the church itself. Sometimes, standing by the villa's fountain, which in those days was still surmounted by a stone carving of a Florentine lily, he must have gazed in frustration across the rooftops of Rome at the dome of Saint Peter's, though with his poor eyesight it would have been something of a blur. Perhaps he knew of Maffeo Barberini's wish to preserve Michelangelo's original design for the façade and now remembered his struggle with Paul V. Surely, he recalled how he and Maffeo had dined together and talked of poetry and philosophy. He must have wondered how his old friendship with the pope had come to this.

That winter, Galileo depended entirely on Francesco Niccolini, but he often disagreed with him too. As ambassador, Niccolini had the habit of cautiously cultivating others' good graces, whereas Galileo grew ever more embattled, more entrenched: his argumentativeness was alarming. Only a few days before leaving Florence, Galileo had written a long letter to his old friend Elia Diodati, an erudite Parisian lawyer, explaining in impassioned terms why Copernicanism did not contradict Scripture: considering that the Bible is couched in terms accessible to simple folk, he asked, "Why must we begin our investigations with God's word rather with His works? . . . God himself is subject to anger, to repentance, to forgetfulness" in the Bible. To take such capriciousness literally would be pure heresy. "But I would ask whether God, to accommodate himself to the capacity and opinions of the same simple folk, has ever changed his creations, or if nature, his inexorable minister, impassive with respect to men's opinions and desires, has not always preserved and maintained the same sort of movements, geometrical figures, and dispositions in the parts of the universe, the moon being always spherical."

Galileo continued to insist upon these views during the course of the winter. And, well, maybe he was right, maybe he was wrong, but as far as Niccolini was concerned, this wasn't the moment. Was he going to raise debating points with the Inquisition, as if this were a seminar at the University of Padua? Niccolini decided it would be wise to take Galileo to see Monsignor Alessandro Boccabella, who had just left his position as assessor at the Holy Office and was amiably disposed toward him; they also inquired after the commissary general, Fra Vincenzo Maculano da Firenzuola, who would likely have more power over Galileo at his trial than anyone but the pope, but Maculano, who would later serve as the papal principality's chief military architect, was out, unfortunately. In the meantime, Cardinal Barberini told Niccolini to keep visitors away from Galileo, and please, the less jabbering the better: nothing he would say was likely to help him. Niccolini agreed: it was best not to rock the boat. "Though the affairs of this tribunal can never be discussed clearly or with any certainty," he wrote Cioli, "it seems that no great evil is upon us yet."

Left to his own devices, Galileo tried to persuade himself that all was going well. He wrote Cioli on February 19 that his case was being handled "according to a very mild and benign principle of treatment, wholly unlike the threatened racks, chains, dungeons, etc." And he went on, in the dressy style he fell into when upset, "To have heard from many, and also to have seen, that there is no lack of persons, and powerful persons too, whose affection for me and my affairs reveals itself to be nothing if not well disposed, is a source of consolation." One notes that the reference to torture is immediately followed by tortured syntax, and also that if the threat of a *quaestio*, a "rigorous examination," had been waived in his case, as certain writers have claimed, no one had told him or Niccolini about it. It had certainly not been waived but would reappear, amid all the apparent official benignity, in the form of a specific pontifical request for the "threat of torture" some four months later.

Over the next few days there was a curious development. A chatty Inquisition official with the title of consultor, a theological position, turned up repeatedly at Villa Medici. His name was Lodovico Serristori, and he came, Niccolini thought, "in the guise of a visitor," as if on his own account. But he always inquired about Galileo's case, prying even into its technicalities, so the ambassador concluded that he had been sent to sound out Galileo—to gauge his rhetorical skills and weigh his defense of his cause—in order to report back to his superiors. In other words, Serristori was a spy, but in this Rome of 1633, that scarcely excluded his being a sympathetic spy, even a spy with divided loyalties. And Galileo, forgetting Cardinal Barberini's advice, poured out his soul to the officious caller, whom it seemed he knew from way back. Together they reviewed Galileo's writings, which Serristori said he admired profusely. "I think Serristori has cheered up the dear old fellow," a bemused Niccolini told Cioli. "And then again it comes back to Galileo, how strange it is, this persecution. I told him, Whatever you are ordered to do, always be ready to submit and obey. Because that is the way to assuage the ferocity of those who are fired up against you, those who treat this cause as something personal."

In their correspondence, Galileo and his friends Niccolini, Cioli, Castelli, and others often refer to this cabal of personal enemies, animated (so they claim) by envy. They do not name them, since their letters, transmitted by courier, are in no way secure. But surely they have specific people in mind, and one wonders whom they refer to. Galileo specifically names his former supporters, the Jesuits. Among them were Christopher Scheiner, the gifted astronomer who had quarreled with Galileo in 1611–13 about the sunspots, though one questions his weight with the Holy Office; Father Orazio Grassi, who bore a long-standing grudge against Galileo after the dispute over the comets; and Melchior Inchofer, a Hungarian-born polemicist who would later file a scathing report against Galileo at the behest of the inquisitors. And of course Galileo, like any important man, had a host of petty critics and belittlers. Yet the great question is whether Niccolini and company weren't really talking about the pope. Clearly Maffeo Barberini had personal reasons to be embittered against Galileo, but he had objective reasons, political and doctrinal, to be worried about him as well. Whether he would have behaved any differently had he not had an axe to grind is a question that will never be resolved.

Curiously, Grand Duke Ferdinand, anxious to intercede on Galileo's behalf, did not turn to the pope but elsewhere. He decided to write two cardinals, Desiderio Scaglia, a respected theologian, and Guido Bentivoglio, a cultivated Ferrarese whose family had for a while owned Palazzo Borghese and who had been painted by Van Dyck, urging an expedited trial for Galileo on account of "the compassion that he deserves" and "the love that I bear him." Bentivoglio politely called on Galileo; Scaglia did not. Yet despite Galileo's conviction that these princes of the Church were now in his corner, the duke's entreaties failed to achieve their desired end.

This may have been because a disconcerting development, in the form of fresh evidence, had arisen. "From what I gather," Niccolini told Cioli in late February, "[Galileo] was given an injunction as early as 1616 against either discussing or teaching the [Copernican] opinion: but he says that the commandment was not at all in this form, but was rather that he neither hold nor defend it." This was perplexing, and Niccolini inquired further. One thing he knew for certain: that if the Inquisition could prove that Galileo had personally received a formal warning of this sort, his conscious infraction of it would trump the subsequent permission he had received to print the *Dialogue*. The Holy Office would have an open-and-shut case.

Here are the circumstances that soon came to light: Everybody knew that in March of 1616, the Congregation of the Index had issued a decree against Nicolaus Copernicus's *Revolutions*, against a letter by the Carmelite monk Antonio Foscarini, and against the *Commentary on Job* by the Spanish priest Diego de Zuñiga, the latter two being pro-Copernican works. The Copernicus and the Zuñiga publications were ordered "suspended until corrected," Foscarini's was suppressed outright. Galileo, though he had explicitly backed Copernicanism in *Letters on the Sunspots*, of 1613, was not mentioned, perhaps out of regard for his patron, the grand duke of Tuscany, but he could hardly have failed to get the message. Of course, since the summer of 1632 the Holy Office had strongly suspected that Galileo had violated this decree in publishing the *Dialogue*—that was the reason for his presence in Rome.

More recently, the Special Commission appointed to investigate the matter had unearthed a memorandum in the Inquisition's archives. It stated that on February 26, 1616, at a meeting with Cardinal Bellarmine, then the commissary general of that body, Galileo had received not only an oral but also a written injunction, similar to a "charitable admonition," ordering him to abandon forthwith any leanings he might have toward Copernicanism. "At the Palace," this memorandum stated, "the usual residence of . . . Cardinal Bellarmine, the said Galileo . . . was . . . admonished by the Cardinal of the error of the aforesaid opinion and that he should abandon it; and later on in the presence of myself [the notary], other witnesses, and the Lord Cardinal, who was still present, the said Commissary did enjoin on the said Galileo, there present, and did order him . . . to relinquish altogether the said opinion, namely, that the sun is the center of the universe and immobile, and that the earth moves; nor henceforth to hold, teach, or defend it in any way, either verbally or in writing. Otherwise proceedings would be taken against him by the Holy Office. The said Galileo acquiesced in this ruling and promised to obey it."

This document shows several technical irregularities, and experts have written much about it, even in the past few decades. But it was accepted as genuine in 1633 and it enraged the pope, who claimed that Galileo had concealed it from him. It put a whole new complexion on the trial.

Galileo's response to this development astonished Niccolini. Indeed it was here that Niccolini's tact and worldliness came to the fore and that the two friends' views on the trial parted absolutely, for some time to come. Strangely, a buoyant Galileo wrote to a friend in Florence that "those numerous and most grave imputations [against me] have all been reduced to one point"—he meant the 1616 written injunction from Bellarmine—"and all the others have ceased; on this basis alone I shall have no trouble securing my release." But Niccolini, confessing to Cioli that he had lost all hope for an expedited trial, suggested that Galileo, if he knew what was

good for him, should proceed with *amorevolezza*, or loving
affection, a curious locution in the circumstances.

Amorevolezza: this word reveals at one stroke the enormous psychological abyss separating Niccolini's and Galileo's
estimation of the perils ahead. And because Niccolini understood the papal court so much better than Galileo, it also
reveals the abyss separating Galileo and the Church. The
emotion of *amorevolezza* is generally reserved for members of
one's own family—it is something you feel for your mother or
your favorite aunt or uncle, or a grandchild who particularly
cares for you. Niccolini's advice to approach the Inquisition
with *amorevolezza*, which to us may sound utterly bizarre,
resulted from his shrewd sense of how that body viewed
Galileo's position. Ever since *Letters on the Sunspots*, Galileo
had wanted to argue as a believer with the Church. He had
wanted to convince the Church that nature (that is, the sun's
true position in the universe) and Scripture were in harmony.
One could line a bookshelf with volumes about the supposed
debate between Galileo and the Church, but it was a debate
that never took place. It never took place because the Church
did not engage in debates with people, especially not with
people who had wandered off in the direction of rash or erroneous philosophical ideas. No: if what Galileo sought was
intellectual engagement, what the Church sought was discipline. The Inquisition's aim was not intellectual but discipli-
nary, in the sense that a parent needs to discipline a wayward
child. And in this sense discipline is ultimately an exercise of
love. The strategist in Niccolini grasped that the inquisitors
saw Galileo as a kind of prodigal son, a famous and gifted
Catholic who must at all costs be reconciled with Church doctrine. So *amorevolezza* was the best tactic to soften them up.

Meantime, to forestall disaster, the ambassador again tried to intercede with the pope. He mentioned Galileo's age, his ill health, his prestige at the Tuscan court. But the pope spoke of how ponderously the Holy Office deliberated and of Galileo's audacity in expressing his opinions. He also intimated darkly that the detested Ciampoli was behind the whole affair (which suggests, interestingly, that he somehow saw a tie-in between Galileo and a pro-Spanish camarilla in the Vatican, with whom Ciampoli had nefarious connections). Worst of all, Galileo had flagrantly disobeyed the written order given him in 1616 by Cardinal Bellarmine.

Niccolini had come up empty-handed, but at least Maffeo hadn't lost his temper—that was something. So Niccolini decided to try the nephew, who was usually so much more sympathetic. And indeed Francesco conceded that "he cared very much for Galileo and admired him as an extraordinary man; but this issue is delicate and might introduce some fantastic dogma into the world and especially into Florence, where wits were subtle and curious." And then wasn't there that little matter of his having argued so much more forcefully in the *Dialogue* for the earth's motion than for its stability? In the end, Niccolini was given to understand that he should be grateful he had obtained Galileo's detention at the Tuscan embassy, instead of the dungeons of Palazzo Pucci or the Castel Sant'Angelo.

Niccolini secured one more audience with the pope before the opening of the trial. The meeting took place on March 13. This time, the ambassador invoked Duke Ferdinand's willingness to assume redoubled indebtedness to the Holy See if the matter could be expedited, but the pope dismissed the offer. "May God forgive [Galileo] for considering these questions,"

Maffeo said, as reported by Niccolini to Cioli, ". . . and may God forgive our little Ciampoli also, since he too is fond of them and is a friend of the new philosophy. Galileo was my friend, and we have conversed and broken bread together several times at my home, and I regret having to displease him, but this bears upon the interests of the faith and religion." It occurred to Niccolini to suggest that if Galileo "might speak for himself he could easily give every satisfaction, with all proper reverence to the Holy Office; but the pope answered that he would be examined in due course, and there is an argument that they have never been able to answer, which is that God is omnipotent and can do anything; and if he is omnipotent, why do we wish to restrict him with contingencies?" This of course was Maffeo's pet thesis. Niccolini, pouncing on the opening, told Maffeo that he had clearly heard Galileo say that "he himself didn't believe in the earth's motion, but since God could make the world in any one of a thousand ways, one couldn't deny that he might have made the world in this one."

Ah, diplomacy.

But the trouble with diplomacy, as Niccolini knew, was that it seldom worked with Maffeo Barberini. As a temporal prince, Maffeo received continuous visits from foreign ambassadors, usually representing Catholic states, and often they came away stunned. The pope did not listen, cutting off his interlocutors; or he listened only to himself, holding forth for hours on end; or he harangued one envoy and then carried the dispute forward to the next audience, causing total bewilderment; or he refused petitions or demolished arguments merely to show off his reasoning powers. The Venetians, with their antipathy toward the papacy, treated him like a spiteful child, requesting the reverse of what they desired in the conviction that he would take the opposite course.

This time, Maffeo's response to Niccolini was annoyance mounting toward ungovernable rage. "One cannot impose necessity on the blessed Lord," he barked. And Niccolini, beginning to worry lest he drift unguardedly into some heretical opinion and end up before a tribunal of the Inquisition himself, passed on to other business.

On April 9, only three days before the trial was to open, Niccolini visited Cardinal Barberini to see whether Galileo might not remain at the Tuscan embassy instead of being confined for the duration of the proceedings in the palace of the Holy Office. Niccolini assured Francesco of the esteem of the Casa Medici, lamenting that he could not hope to paint a suitably vivid picture of the "failing health of the dear old fellow, who for two nights together has cried out and complained continuously of his arthritic pains and of his wretched old age and the suffering it causes him"—all good reasons to permit Galileo to return each night to sleep at the embassy. Francesco regretted his incapacity to make such a concession, but he promised to arrange comfortable lodgings for the scientist in the chambers of the Inquisition's treasury. There Galileo would be free to come and go, to take exercise in the courtyard, mingle among the churchmen, and receive his meals from the kitchen of the Villa Medici.

Meanwhile, Galileo prepared to defend his philosophical position, much to Niccolini's dismay. Convinced as always that fortune favored him, he appeared to have some polemic or piece of evidence up his sleeve that would sink his persecutors' arguments. The ambassador pleaded with him to abandon this project and to toe whatever line the Holy Office might lay down, especially with regard to "that particular" of the earth's motion; but he soon saw that such thoughts depressed Galileo so deeply he began to fear again for his life. The great man deserved all the best, Niccolini felt; "this whole house," he told Cioli, "which loves him exceedingly, feels an indescribable pain."

The trial of Galileo Galilei opened on April 12, 1633.

"*HOW AND IN* what season did you arrive in Rome?"

"I arrived in Rome the first Sunday of Lent, and I came in a litter."

"Did you come of your own will, or did you receive a summons from certain persons that you come to the city, and to what end?"

"At Florence the Father Inquisitor ordered me to come to Rome and present myself before the Holy Office, this being the command of the ministers of this Holy Office."

"Do you know or can you imagine the reason for which you have been enjoined to come to Rome?"

"I imagine that the cause for which I was ordered to present myself before the Holy Office in Rome was to give an account of my recently printed book."

With these words the trial opened in an upstairs chamber of the magnificent Dominican convent of Santa Maria sopra Minerva, near the Pantheon. Galileo was interrogated by Friar Carlo Sinceri, the proctor fiscal, almost certainly in Italian, although the questions were entered into the record in Latin; his answers, noted in Italian, sound very much like Galileo's own words. The mental picture that most of us have of this notorious event is an amalgam of many impressions: the memory of old paintings and engravings; perhaps Bertolt

Brecht's effective but factually inaccurate play; surely a strong sense of what was historically at stake; and, I would argue, an unsuspected residue of Perry Mason. I say Perry Mason because we in the English-speaking world have come to see a trial, any trial, as an occasion in which two sides contend, each presenting evidence on behalf of a narrative model held to be essentially valid. So we assume that the dispute in Rome in 1633 concerned the nature of the solar system and the accuracy of assertions about it. We know that Galileo was right about that, and we know that his accusers were wrong, and this gives us our vision of the trial. A row of pinch-faced, black-clad priests in funny hats sit facing an elderly, rather stooped, but nobly defiant scientist, a good Catholic as it happens, who raises his arm in a demonstration of a physical principle. In response, the judges quote scriptural chapter and verse, until at last the old man is silenced. As he leaves the courtroom he mutters, *"Eppur si muove"*—and yet it *does* move—unable to betray the truth.

Of course, this picture (and, unhappily, Galileo's parting line, of which there is no record in the transcript) does not correspond to reality. For one thing, the idea of an adversarial trial would not have seemed natural to the cardinals of the Inquisition. For another, most of them (though not all) were products of the Renaissance, sophisticated humanists as well as post-Tridentine prelates, and God only knows what inner reservations some of them may have had about the proceedings. We cannot know their secret thoughts, or enough about this trial in general, but one thing we do know is that like all such trials it was "inquisitorial" (this time with a small "i") in its juridical form. This means that no Perry Mason–style debating took place, but rather the Commission of Inquisitors balanced the evidence for and against the accused, as is still the case in certain Continental judicial systems. In accordance with this procedure, Galileo was deposed as needed, being formally addressed in Latin, and permitted to answer in Italian, with his statements recorded by a clerk.

The indictment was arrived at during the course of the proceedings, which strikes us now as outlandish, but it raised no eyebrows at the time. The central accusation—whether Galileo was guilty of "temerity," that is, rashness (contradicting the Church fathers' interpretation of Scripture), or out-and-out heresy, a violation of the faith—was to be decided on the basis of the evidence obtained as the trial unfolded. Galileo's astronomical ideas were pondered only once, in a short, dismissive paragraph written by a churchman without scientific qualifications, and no expert witnesses in the matter were called. However, theologians were enlisted to comb his writings for signs of a positive endorsement of the heliocentric opinion.

If the Vatican's extant transcript of the Galileo trial is complete, and it does appear to be, the proceedings show signs of a departure from standard practice. The Roman Inquisition has hardly been studied, in part because its archives were closed to scholars until 1998, and we know little of the general run of cases in Galileo's day. Yet the scrutiny of canon law, legal manuals for inquisitors' use, and a body of records preserved at Trinity College, Dublin, have given us some sense of Inquisitorial practice in early-sixteenth-century Italy.

The Roman Inquisition was founded by the papal bull *Licet ab initio* of 1542 as a response to the Protestant threat in the peninsula, and a commission of ten cardinals, later called the Congregation of the Holy Office, headed by a prosecutor called the commissary general, was set up to administer it. The task of the Congregation, which was always understood to be under the direct authority of the pope, was primarily to pursue heresy, usually Lutheranism, and this mission must be borne in mind. The Inquisition was unquestionably a form of mind police. But because it was a Christian mind police, it aimed not only to discover what the accused was thinking but also to reform his thoughts, that is, to return him, fit for salvation, to the body of the church militant. In the business of saving souls, it was more lenient than any Italian civil authority of the same period. Our chamber-of-horrors vision of the Roman Inquisition is derived largely from tales of the earlier, medieval Inquisition, or from the Spanish Inquisition, or from pulp fiction penned in the Protestant world. Whereas civil courts concerned themselves overwhelmingly with what defendants had done, and were often ready to hang them for it, the courts of the Roman Inquisition wanted to know what they were thinking, and so were inclined simply to imprison them until they changed their mind: in fact, the ecclesiastical courts were the first in Europe to use prison terms as a form of punishment and not only as a means of pretrial detention. This fact is sometimes cited in mitigation of the Inquisition's proto-totalitarian nature, but really it cuts two ways. The civil courts after all did not care what you thought, only what you did, whereas if the Inquisition got wind of any public doubts you harbored about the Virgin birth, you would be in very hot water indeed. By the same token, the Galileo affair came about not, as one might imagine, because Galileo had one view and the Church another, but because Galileo had a certain view and the Church insisted that he change it.

Since the Inquisition cared so very much about what the accused was thinking, he was generally permitted to read the articles of his indictment and to prepare a defense. Theoretically, he could also study the evidence against him and rebut the testimony of his accusers, call his own witnesses, and enlist the aid of a court-appointed attorney well-versed in canon law (though not in a manner resembling current American or British practice—he had, for example, no opportunity to question the legal merit of the accusation or the court's jurisdiction, no right to cross-examine witnesses, no in-court litigator, and no access to a court of appeals). If, in the end, he threw himself on the mercy of the tribunal, his petition might be favorably considered, since a change of heart was the desired outcome. Nowadays we assume, in accordance with Article Ten of the Universal Declaration of Human Rights, that a tribunal must be "independent and impartial," that the accused must know exactly who is accusing him, that a trial must be public, and that the jury must be composed of one's peers. Galileo enjoyed none of these rights. A defendant moreover is to be presumed innocent, yet in the more heavily consulted jurisprudential manuals of this period, the issue of the presumption of innocence or guilt is not unambiguously posed. To sum up, if the evidence against Galileo was thoughtfully weighed in the spring of 1633, and he was permitted to speak in his defense, we should not imagine that his hearing resembled a trial in our sense of the word.

The *quaestio*, or judicial torture, a holdover from the medieval Inquisition, was applied above all in heresy cases. The culprit, it was felt, did not have the right to withhold evidence about himself or his accomplices or to issue an insincere confession. To attain this information, which the court con-

versely had the right to know, a limited amount of physical torment could be applied, and the accused, whether determined
to be guilty or innocent, was to suffer no permanent harm by
the end of his ordeal. Because of these restrictions, and because
the instrument used, the *corda*,* allowed for the incremental
application of pain, there is little difference between the Inquisition's prescription of torture and that practiced by modern
Western governments, including the United States. The aim
was and remains to discover some secret intention by means of
a measurable administration of pain that does not—
theoretically—leave any permanent damage.

The Inquisition's use of torture was curtailed by numerous
cautionary restrictions, some of which bear directly on the
Galileo case. Ideally, both the bishop and the chief inquisitor
in question had to be present. The exercise was not to last
more than an hour at most, and clerics, the gentry, and persons of unimpeachable character were exempt, as were those
already convicted, children, the old or infirm, and those who
had eaten within six to ten hours, lest they vomit. "Torture,"
went a canon-law maxim, "is a fragile and dangerous thing,
and the truth is frequently not obtained by it. For many defendants because of their patience and strength are able to spurn
their torments, while others would rather lie than bear them,
unfairly incriminating themselves and also others." Sometimes indeed the accused bore torture without altering anything in his original testimony, and if he denied harboring

* The corda was a mechanical device by which the culprit's arms were
bound behind his back and he was lifted into the air by a rope looped
through a pulley. Pain was applied by dropping him repeatedly and
abruptly to within several inches of the floor.

heretical beliefs, his ordeal might be considered proof of his innocence and a prelude to absolution. In any event, confessions obtained by such means were not considered valid until ratified by the accused at least twenty-four hours later, outside the torture chamber.

The most cursory review of these canon-law guidelines reveals that they were not followed in the Galileo trial. Galileo was never given the right to have a lawyer, to prepare a thorough defense, or to call friendly witnesses. Moreover, as his letter of February 19 to Benedetto Castelli reveals, he felt himself under threat of torture, or to put it more mildly, no one had seen fit to relieve him of this perhaps unwarranted apprehension. (As we will see, the trial transcript for June 16, 1633, holds a clear pontifical order to interrogate Galileo *super intentione*, "concerning his secret intentions," *etiam comminata ei tortura*, "also under the threat of torture.") Yet at almost seventy he was too old to sustain the *corda*. He was manifestly unwell, and his character, if testy, was irreproachable. And finally, curiously enough, he was technically a cleric, having been tonsured on April 5, 1631, by Monsignor Strozzi, the archbishop of Siena, in order to receive a church pension.

One can only conclude that canon law was not properly followed during the Galileo trial or, less plausibly, that it was being generally applied at that period in a manner for which our contemporary legal historians have found no substantial precedent.

There is another matter that we ought to consider before we examine the course of the trial. Very often, when a question bearing heavily on the public interest is decided by a high or influential court, the popular press automatically assumes that the issue itself has been substantively adjudicated. When a closer look at court documents discloses that the question has been resolved on the basis of some point of law such as jurisdiction or untimeliness, the pundits speak of "technicalities" or "loopholes." Of course they are nothing of the sort for lawyers. In the Galileo case, analogously, two substantive issues were theoretically at stake: whether the Church might accept the Copernican hypothesis, and whether it might countenance, with respect to cosmology, a layman's figurative reading of some verses of the Bible. In reality, however, the issues to be resolved were precisely what we might call technicalities: whether Galileo had a proper license to print the *Dialogue*; whether he had disobeyed Cardinal Bellarmine's 1616 injunction not to teach Copernicanism; and whether his discussion of Copernicanism was hypothetical or conclusive. The truth is that just as no churchman had been authorized by the Vatican to challenge Galileo's heliocentrism before the trial (though several did, very weakly), none seriously challenged it during the trial. The reason is obvious. It contradicted Scripture. It was a nonissue.

If there is little interest in the Galileo trial from an intellectual point of view, however, there is much interest from other points of view. One aspect, not often considered, is the psychological. For three months Francesco Niccolini, with his diplomat's insight into the papal court, had been coaching the fiercely resistant Galileo as a future deponent, trying to get him to abandon his argumentativeness and egocentrism, trying to get him to submit. Thus, when Galileo appeared in court, the language of his depositions and his replies to the magistrates sometimes seems strangely binary, as though a small, rebellious defendant were hiding inside a larger, more compliant one.

On the other side, the inquisitors were hostage to the Vatican's own equivocation, an ambivalence that went back to the decree of 1616, which only "suspended" Copernicus's *Revolutions*, pending certain emendations, and failed to censure Galileo, whom Bellarmine had merely admonished not to advocate the heliocentric opinion. The equivocation had been renewed, even cemented, with Maffeo Barberini's accession to the papacy in 1624 and his promotion of Galileo's friends Ciampoli and Riccardi to positions of authority, all of which had given Galileo hope that the 1616 ban might be revoked. Why had Prince Cesi of the Academy of Lynxes, that avowed foe of the Jesuits, found such favor with Urban VIII? Why had Galileo been encouraged to plug away for a decade at his pro-Copernican dialogue? Why had Riccardi allowed it to be printed, if the Vatican was so opposed to its contents? The answer to these questions cannot be found in the realm of ideas, but rather (if at all) in that of psychology, in the rivalry of personalities and parties. The equivocation, which arguably resulted from the papacy's desire to play two conflicting roles simultaneously—on the one hand, promoter of Tridentine orthodoxy; on the other, renewer of humanism and patron of a Tuscan client-state—was a terrible trap for Galileo, but it was a trap for the Holy Office as well, for that body now found itself in the derisory position of considering whether to suppress a work that bore its own imprimatur.

Galileo at his first deposition must have seemed remarkably optimistic. Under interrogation he began to walk very boldly on very thin ice. He stated that he had come to Rome in 1616, and again in 1624, with the express intention of "making sure of holding nothing but holy and Catholic opinions"; actually, he had hoped to induce the Roman Church to become the sponsor of his own opinions. After conversing with several cardinals, he said, he had discovered that the Vatican regarded the Copernican system as "repugnant to the Holy Scriptures, and admissible only *ex suppositione*, in the manner in which Copernicus himself took it"*; actually, neither Copernicus nor Galileo had regarded the Copernican system as anything but a likely hypothesis in the modern sense. This sort of suppositional reasoning was exactly what Galileo had demolished with the telescope and physical astronomy. Now he was invoking it for the purposes of self-exoneration.

But Galileo wasn't through yet, for a piece of surprise evidence was about to make a fairy-queen appearance. Ever certain that fortune favored his cause, he held up a document for the court to see: one imagines the inquisitors bending forward to peer at it. So *this*, during those long, hard weeks of winter confinement, was what had mysteriously cheered him and convinced him of his impunity.

"I was notified," he began, "of the decision of the Congregation of the Index, and the one who notified me was Cardinal Bellarmine. . . . In February of 1616, Cardinal Bellarmine told me that to be of the Copernican opinion was absolutely contrary to Holy Scriptures—one could neither hold nor defend it—but suppositionally one could take it up and make use of it. In conformity with which I have here a certificate from the same Cardinal Bellarmine, executed on the 26 of May, 1616, in which he says that the opinion of Copernicus

---

* The editor of Copernicus's *On the Revolutions*, Andreas Oseander, had supplied an unapproved and not particularly believable preface stating that the work was suppositional, and of use primarily for mathematical computations.

ld nor defended, being against the Holy Scrip-
a certificate I hereby offer a copy, and here it is."
court gathered was that Galileo, outraged by
e effect that he had been censured by the Holy
Office, had in 1616 sought further contact with Bellarmine of
his own free will, and had received a written clarification and
attestation. Bellarmine had died in 1621, but Galileo's copy of
the document, twelve lines long, was read aloud, marked with
the letter "B," and dutifully entered into evidence:

> We, Robert Cardinal Bellarmine, hearing that it has been
> calumniously rumored that Galileo Galilei has abjured in
> our hands and also has been given a salutary penance, and
> being requested to state the truth with regard to this,
> declare that this man Galileo has not abjured, either in our
> hands or in the hands of any other person here in Rome, or
> anywhere else as far as we know, any doctrine or opinion
> which he has held; nor has any salutary or any other kind of
> penance been given to him. Only the declaration made by
> the Holy Father and published by the Sacred Congregation
> of the Index has been revealed to him, which states that the
> doctrine of Copernicus, that the earth moves around the
> sun and that the sun is stationary in the center of the uni-
> verse and does not move east to west, is contrary to Holy
> Scripture and therefore cannot be defended or held. In wit-
> ness whereof we have written and signed this letter with
> our hand on this twenty-sixth day of May, 1616.

When pressed further by the interrogator, who must have
sounded not a little surprised, Galileo described the circumstances of the meeting and the possible presence of certain

clerics. He remembered that perhaps he had also been instructed "not to teach" the heliocentric system, as well as not to hold or defend it. "I don't remember any particular *quovis modo*," he slyly added (referring to a Latin expression in Bellarmine's earlier warning to him meaning "in any manner whatsoever"), "but I didn't think about it or recall it, because several months later I received this certificate from Cardinal Bellarmine." In other words, the letter from Bellarmine, which did *not* exclude hypothetical discussions of the heliocentric position, seemed to override the more sweeping prohibitions of other injunctions, which, besides, he had not remembered.* "After receiving the above-mentioned precept," he went on, "I did not request permission to write the book in question . . . because I do not believe that in writing it I would transgress in any way against the precept made to me, not to hold nor defend nor teach the said opinion, indeed to refute it."

Everything had gone swimmingly for the defendant until the last few words: *indeed to refute it*. There he had gravely erred. As we know from detective stories, superfluous compliance can be self-incriminating: consider the witness who offers more testimony than anyone asked for. Galileo had now done something analogous. Perhaps he might have convinced the Inquisition that he had explained the Copernican system as a sort of philosophical exercise, but did he really think he could convince it that the *Dialogue* humbly obeyed an ecclesiastical command to *refute* that system? The Inquisition had been

* He may have received a further warning from a priest named Michelangelo Seghizzi de Lauda, but if this happened, he apparently forgot about it as well.

studying the *Dialogue* for eight months, so Galileo's preposterous claim can only have suggested that he was speaking insincerely, telling the interrogator whatever he wanted to hear. It may be that he had taken Niccolini's counsel too much to heart in the end. Unfair trials, however, have an extraordinary capacity to induce otherwise honest people to lie: the trial's unfairness and Galileo's lying must be taken as correlatives.

If, then, for Galileo, Bellarmine's certificate was a ticket to exoneration, the Inquisition must have begun to see it very differently. "Malice" and "deception" were what the tribunal most abhorred. Remember that the culprit needed to explain why he had not told Riccardi, the Vatican censor, of Bellarmine's earlier warning. The reason, he intimated, was that it had been clarified and superseded by this document. But if one listened carefully to Galileo, here he was claiming that his *Dialogue* aimed to "refute" the Copernican system, which sounded like simple perjury. Moreover, Bellarmine's certificate had itself clearly admonished him neither to defend nor to hold the Copernican "opinion." Had he, in fact, "held" it? Soon a panel of three theological consultors would issue their reports. Suppose the *Dialogue* should indeed be found to defend or to hold Copernicanism (not to mention teaching it or discussing it, both of which Bellarmine had forbidden); wouldn't that provide a perfect justification for Riccardi and the entire Holy Office? Instead of being in the doubtful position of banning a book they had licensed, they could now claim that Galileo was maliciously availing himself of Bellarmine's certificate in order to justify obtaining permission for a volume injurious to the faith.

So the interrogator logically turned to Galileo's request to publish the book. Galileo recalled his trip to Rome in 1630 to obtain an imprimatur from Riccardi. He cited his reception of it, on Riccardi's condition that he, Riccardi, be allowed to "add, delete, and change as he saw fit" before publication, which had led to the pope's brief and clearly inadequate review of the *Dialogue* and to the insertion of its contradictory preface and conclusion (an echo of the pope's own theology, which actually contradicted the book's thesis and were voiced by the callow Aristotelian, Simplicio). But upon Galileo's return to Florence, the matter had been suspended, as Riccardi fretted over the book's contentions. Meanwhile the Black Plague broke out, rendering travel hazardous. Riccardi agreed to turn the request over to the inquisitor in the Vatican's Florence office, as long as the preface and conclusion be submitted to him for final approval. So it had come about, Galileo said, that the *Dialogue* had been printed in Florence, with no less than one imprimatur from Riccardi and another from the Florentine inquisitor (and—though he didn't mention this—the preface and conclusion in an incongruous typeface).

This account left him open to an obvious question. When he had asked Riccardi for permission to print the book, the interrogator asked, had he revealed the injunction previously given to him by the Inquisition?

"I did not," Galileo replied, ". . . because I did not judge it necessary to tell him, since with this book I had neither held nor defended the opinion of the earth's motion and the sun's stability."

This was nonsense. For he had held and defended it, if only (as he claimed) suppositionally; though, when you came right down to it, he had held and defended it conclusively and affirmatively. After one short hearing, Galileo stood on the verge of losing his freedom forever.

He also stood on the verge of winning it. The Tuscans had a certain clout in Rome, and Niccolini, for one, had been active behind the scenes. He had asked the grand duke to draft pleas to all the inquisitors not already canvassed. Soon Niccolini received a batch of missives, which he attempted to deliver to them, with scant luck in some instances. He also became aware that his fellow-Lyncean Francesco Barberini and, of all people, the chief inquisitor, Vincenzo Maculano, known as the commissary general, were now laboring to get the trial expedited, exactly as Niccolini had wished from the start. Maculano, it has been suggested, felt his case undercut by the appearance of Bellarmine's certificate. But he was not analogous to an American district attorney—his "career" was not on the line. He was charged not with "winning" the case and securing the maximum penalty, which would have meant seeing Galileo charged with heresy and imprisoned, but with securing Galileo's confession and renunciation of Copernicanism, if it should be proved that he supported it; the book could be condemned or emended as needed. One may speculate that the authority of Bellarmine's certificate might only have softened Maculano's already malleable position. On April 16, Niccolini wrote Cioli that Maculano had told Niccolini's secretary that the trial would soon be concluded; indeed, Maculano had informed Niccolini himself that Francesco Barberini had been pleading the scientist's case to his uncle, trying to "palliate" the pope's "emotions." He wrote Cioli a jubilant note on April 23, declaring that Galileo would probably be released on or right after Ascension Day, when Urban VIII returned from his country residence of Castel Gandolfo.

What had happened? The day before, in a letter discovered only in 1999, a distraught Maculano had written the following words to Francesco Barberini: "Last night Galileo was afflicted with pains which assaulted him, and he cried out again this morning. I have visited him twice, and he had received more medicine. This makes me think that his case should be expedited very quickly." Recently, he said, the three theological consultors' reports had been filed, all of which concluded that Galileo "defends and teaches the opinion which is rejected and condemned by the Church, and that the author also makes himself suspect of holding it. That being so, the case could immediately be brought to a prompt settlement." In another letter to Francesco, Maculano reported on April 23 that he had proposed a plan for a settlement to the Holy Office, requesting that the cardinals "grant me the authority to deal extrajudicially with Galileo, in order to make him understand his error and, once having recognized it, to bring him to confess it. The proposal seemed too bold at first . . . however, after I mentioned the basis on which I proposed this, they gave me the authority . . . In order not to lose time," Maculano continued, which suggests that as a clergyman with a conscience he was extremely worried about Galileo's health and mindful of canon-law restrictions about trying defendants *in extremis*, "yesterday afternoon I had a discussion with Galileo, and, after exchanging innumerable arguments and answers, by the grace of the Lord I accomplished my purpose: I made him grasp his error, so that he clearly recognized that he had erred and gone too far in his book; he expressed everything with heartfelt words, as if he were relieved by the knowledge of his error; and he was ready for a judicial confession. However, he asked me for a little time to think about the way to render his confession honest. [Thus] the Tribunal will maintain its reputation; the culprit can be treated with benignity; and, whatever, the final outcome, he will know the favor done to him, with all the consequent satisfaction one wants in this."

One cannot resist picturing the two Tuscans arguing in the waning afternoon light, the humane and practical-minded priest sitting erect in his crimson vestments, now and then bending to help the half-reclining scientist ease his pain. The high point of the drama had been reached: one sees Galileo falling into his irrepressible habit of talking mathematics and biblical hermeneutic, and Maculano as a priest (and an engineer, with mathematical training) tactfully warning him off this course. One can only speculate about Maculano's reasons for wishing to strike a deal with Galileo, but certain possibilities (in addition to the latter's ill health) come to mind: the pope's own hesitancy and former friendship with Galileo; the bad publicity generated by the trial in Tuscany, Italy, and Europe at large; the obvious weaknesses in the prosecution; and the probability that, given Galileo's reputation and the widespread interest in Copernicanism among Italian intellectuals and even many churchmen, his massive condemnation and punishment would constitute very poor public policy. Finally, one cannot exclude a truant sympathy for the man and his ideas: Benedetto Castelli, now mathematician to the pope, had written Galileo on October 2, 1632, that Maculano himself (who as a fortifications expert was probably the only member of the Congregation capable of following the demonstrations in the *Dialogue*) "was of the same [heliocentric] opinion, that the question should not be concluded with the authority of the Sacred Letters [i.e., Scripture]; and he even told me that he wanted to write about it." Aside from the foregoing, there must have been, as always in the secrecy-bound

Rome of 1633, a number of hidden political reasons for the plea bargain that we cannot remotely imagine. It would be naive to think that Maculano had no personal or familial motive for ending the trial posthaste.

As for Galileo, in this face-to-face meeting with a prelate of Maculano's standing, he must suddenly have realized that his soul was in mortal danger. On one level, he could not rid his mind of a world picture that he knew to be true. On another, he was doubtless saddened, though perhaps for the first time, to have so displeased his Church. Caught between two contradictory emotions, and doubtless ashamed of his recent perjury, he surely needed the interval he sought to arrive at what he called an "honest confession." Eventually, under the impress of fear, genuine piety, and a desire to espouse Copernicanism while simultaneously rejecting it, Galileo's duplicity probably turned into self-delusion.

At this point, Latin Christendom hung in the balance. Let us for a moment play a parlor game, the game of counterfactual fantasy. Had Maculano's scheme been realized, Galileo would probably have confessed his error. He would have been convicted of some lesser crime, such as "rashness," and been obliged, most likely, to do penance for several months. He would have been permitted to return to his research with full freedom of movement as long as he left Copernicanism and the Bible alone. But by this time Maculano, and perhaps Niccolini too, must have grown aware that not all the inquisitors of the Holy Congregation were of their opinion. Let us continue our parlor game and speculate about the two camps into which they conceivably had separated. In the first, inclining to leniency, were surely Maculano and Francesco Barberini; and, we might guess, Gaspare Borgia and Laudivio Zacchia (because they, like Francesco, would decline to present themselves at Galileo's eventual recantation); also, just possibly, Antonio Barberini, the pope's brother, who had been shoehorned into the curia without qualifications nine years earlier and who, according to Niccolini, liked to be wooed, since it made him feel important: Niccolini told Cioli that Antonio Barberini might well be "helping" Galileo more than anybody else, and what could this mean but badgering the pope? So we might suppose (without, I repeat, any evidence) that these inquisitors were "soft," and the others presumably less so or—unhappily—not at all so.

Two days after Maculano's fateful meeting with Galileo, the accused gave his second deposition. His story had completely changed. No longer did it seem to him that his book refuted Copernicanism. It had now "dawned" on him, he said, to review the book, and he had "started to read it with the greatest concentration and to examine it in the most detailed manner. Not having seen it for so long, I found it almost a new book by another author." He felt that the Aristotelian side in the dialogue had not been presented fairly, and that in trying to show off his polemical abilities in defending the weaker, Copernican side, he had exaggerated. "I resorted," he said, ". . . to the natural gratification everyone feels for his own subtleties and for showing himself to be cleverer than the average man, by finding ingenious and apparent considerations even in favor of false propositions. . . . My error then was, and I confess it, one of vain ambition, pure ignorance, and inadvertence."

In essence, Galileo was being so bold as to tell the tribunal to convict him of vanity, ignorance, and carelessness, none of which approached heresy. The court adjourned briefly, after which he returned to confirm his denial of the heliocentric thesis; he also offered to add, by the tribunal's leave, two more literary "days" to the *Dialogue*'s four, in which his characters would demolish the arguments already presented. The cardinals of the Holy Congregation did not seem thrilled by this suggestion.

SO GALILEO HAD made his confession. He had bowed his head. Niccolini, Cardinal Barberini, Maculano, and anyone else who had discovered what was going forward awaited a rapid resolution of the trial. Yet nothing had actually been determined. There seemed to be some mysterious holdup.

Why had Galileo not been convicted, scolded, given a penance, and released? Unknown to him, after the trial's second session a large amount of material from the Inquisition's 1615–16 investigation of his beliefs, itself notably sloppy, had been entered into evidence as an attachment to a remarkably mendacious summary—not a transcript—of the proceedings to date; this joined the negative reports on the *Dialogue* filed on April 27 by the three theological consultors, none of whom was proficient in astronomy but all of whom concluded that Galileo supported heliocentrism. It is impossible to say who compiled the falsified summary, bloated with the 1615–16 allegations, or to what degree it may have misled the magistrates; after all, they had read Galileo's own account and taken note of his statement of contrition. The three consultors' reports were distinctly damaging, but Maculano had told Niccolini that they strengthened, not weakened, his case for a plea deal. Unfortunately, however, both the summary and one of the reports, a vitriolic attack by the Hungarian-born Jesuit named Melchior Inchofer, who mentioned Galileo's dispute over the sunspots with his still-aggrieved fellow Jesuit

Christopher Scheiner, cited a certain passionate letter that Galileo had written in 1614 to Benedetto Castelli, then a mathematics professor at Pisa. The letter was a private communication to a Benedictine priest; it did not publicly teach or advocate anything; moreover, the Inquisition's copy had been shamelessly doctored. But unfortunately it revealed its author's genuine belief that the Old Testament passages appearing to contradict science should not be taken literally. This was news to nobody, but it may have muddied the case for a plea deal, because the Index had established the contrary in 1616 and because Galileo as a layman had no right to pronounce on the Bible.

It is impossible to deduce from the trial records (which, spotty as they are, may have been typical for that period and that institution) whether Cardinal Maculano's extrajudicial settlement was brutally sabotaged or simply delayed, watered down, denatured through bureaucratic manipulation. In the end, it never went through, and there is no acknowledgment in the documents that it didn't or an explanation as to why it didn't. We do not know what Maculano or Francesco Barberini or their allies made of this shipwreck, but it is not unreasonable to suppose that they grew aware that their antagonists had outfoxed them. They were men of the cloth, however, and they left no lament.

Galileo had been escorted back to Villa Medici on April 30 and instructed to wait. Never had his perception of other people, indistinct at best, been as fuzzy as it was now. To his friends in Tuscany he wrote buoyant letters suggesting that he would be released in short order, and, indeed, exculpated. Niccolini for his part seemed fairly heartbroken. The grand duke, proprietor of the enormous villas of Palazzo Pitti,

Careggi, Cafaggiolo, Poggio a Caiano, Poggio Imperiale, L'Ambrogiana, Serravezza—one could go on and on—was still fussing about the defraying of Galileo's expenses, a detail over which the ambassador scarcely had the will to bicker. Ever since autumn, even before Galileo's arrival, he had been trying to swing an expedited trial, enlisting all his political savvy, all his knowledge of the papal court, and now it was slipping through his fingers. In late May, when the situation had drastically deteriorated, he was still keeping the truth from the old fellow, who was often bedridden. "I haven't told him everything yet," he wrote Cioli, "because I intend, in order to avoid causing him pain, to go at it little by little."

On May 10, Maculano summoned Galileo before the Congregation and offered him eight days to prepare a defense. He coolly replied that he would hand in a written defense forthwith, together with the original of Bellarmine's certificate. Galileo's defense, what we would call a "sworn statement," stated that the Copernicanism of the *Dialogue* was inadvertent and that he was prepared to make amends for it, and he begged the judges to consider his ill health and the slander he had suffered. The Italians have a charming term for an insincere confession of this sort, that it is *figlia della convenienza processuale*, a "daughter of trial strategy," yet it only reflected the plea that Galileo had contrived under Maculano's supervision. Unknown to him, however, the deal was off, and the plea fell on deaf ears.

Someone, or some people, somewhere, had decided to make an example of Galileo. Pope Urban VIII had ultimate authority over the Congregation, and we shall come to this preeminent fact; certain Jesuits may have thrown their weight against the Tuscan physicist; and he had plenty of other enemies. What should be noted, however, is that the conflict between heliocentrism and religion had made only the most garbled appearance at the trial; no scientific knowledge or reasoning had been put to use. At all events, a brief and tendentious report was submitted to the pope, and a pontifical order was issued prohibiting further publication of the *Dialogue* and ordering that Galileo be interrogated concerning his real or secret intentions, under threat of torture if necessary (*interrogandum esse super intentione, etiam comminata ei tortura*), the assumption being that he might be acting with malice or deception.

Canon law barred the torture of Galileo, and there is no evidence that he was tortured. As Maurice Finocchiaro has emphasized, however, the salient fact is that Galileo's final interrogation, on June 21, 1633, was conducted under the threat of torture, and this gives it a peculiar psychological dimension. One must distinguish carefully between the institutional justification for torture and the real, psychic reason why it might be resorted to or threatened. Torture is widely practiced in most countries, but its adoption and abandonment (at least nominal) in the postwar period by such Western nations as France, Great Britain, Greece, Argentina, and Israel, and its current use by the United States, have led to systematic study. It is very difficult for a person or group possessing physical power over a defenseless antagonist to resist using it, or threatening to use it, and this explains the practice of torture much more convincingly than does the flimsy hope of extorting secret information of any practical value. The Roman Inquisition's relatively restrained use of torture, compared to that of the civil authorities, has been cited in mitigation of its horrors. Yet this argument cuts two ways, as stated earlier, for its appearance in a religious context is precisely what excites our repugnance. If, ultimately, the pope resorted only symbolically to the threat of force to resolve his dispute with Galileo, the symbol—the *corda*—was absolutely central, as the popular imagination has long and justly observed.

At the end of the final session the interrogator asked, "Has anything occurred to him that he wishes to say?"

"I have nothing whatever to say," Galileo replied, usually so voluble.

And what does he hold, concerning the sun being at the center of the world?

"Assured by the prudence of my superiors, all ambiguity within me has ceased," he said, adding that he accepted the stability of the earth.

And with respect to the opinion expressed in his book?

"I conclude that within me I do not hold nor have I held, according to the determination of my superiors, the damned opinion."

At this point he was formally warned that he was nonetheless under suspicion of still harboring the Copernican opinion, and that unless he "resolved to make the truth known," the "remedies of the law" and "opportune measures" would be used against him. This again referred to the "rigorous examination" of torture.

"I do not hold nor have held this opinion of Copernicus," he said, "after I was notified by the injunction that I was to abandon it. Besides, I am here in your hands, you may do as you please."

This last phrase, which can also be translated "I am here in their hands, let them do what they want," gives away Galileo's true state of mind. Read either way, it is a rude, peremptory way of addressing a tribunal of cardinals, especially in 1633. At last he has realized that the plea bargain has evaporated: the rebellious, frightened prisoner who for weeks has been hiding inside the penitent culprit devised by Maculano has, with these exasperated words, climbed out. Everything in this interrogation discloses Galileo's terminal acknowledgment of the brute fact that this trial is really about nothing but discipline and obedience, strength and weakness, menace and fear. Close scrutiny of every one of his answers at the final interrogatory shows that he embraces the doctrine of the Church only because he has been told to, and for no other reason. *I have nothing to say . . . the prudence of my superiors . . . I am in your hands, do as you please.*

"And he was told to tell the truth," reads the transcript, "otherwise recourse would be had to torture."

"I am here to show obedience," said Galileo, "and I have not held this opinion after the determination was made, as I have said."

IN THE INTERESTS of following the story of the trial as directly as possible, we have not paused over one circumstance that gravely complicated Galileo's case. This was the appearance, among the documents entered into evidence, of the letter he had written early in 1614 to Benedetto Castelli, in which he had trespassed into the field of biblical exegesis. The letter, as already noted, was private, but once it became public Galileo faced what Richard J. Blackwell has called a kind of "double jeopardy," one point of which related "to the content of the interpretation, the other to assuming the role of being an interpreter. No matter what the merits of the former, the individual was always in jeopardy on the latter." In other words, whether or not the biblical analysis found clerical acceptance, and in this case it did not, Galileo as a layman had no dispensation to interpret the Scriptures. This prohibition had figured among the decrees of the Fourth Council of Trent, published in April of 1546.

What was in this letter, and why did Galileo write it? At the end of 1613, Dowager Duchess Cristina of Lorraine, mother of Cosimo de' Medici, had raised the issue of the compatibility of Catholicism and Copernicanism at a breakfast that Benedetto Castelli had attended. Informed of this, Galileo resolved to draft a short essay expressing his views on the matter, which became known as the "Letter to Benedetto Castelli." He must have intended for this letter to reach some influential people beyond Castelli and so perhaps to forfeit its private character, but he could hardly have suspected that it would end up as evidence for a heresy charge against him. His aim was simply to show that the heliocentric world view did not contradict Scripture and thus posed no threat to the Church. With reference to the best-known scriptural passages that seemed to support the geocentric cosmology, such as Ecclesiastes 1:5 ("The sun rises, and sets, and returns to its place, from which, reborn, it revolves through the meridian, and is curved toward the North") and Joshua 10:12 ("Sun stand thou still . . ."), he argued that the writers of the Bible had accommodated their language to the grasp of ordinary people, offering them the historical chronicles and ethical precepts needed for their salvation; it was perfectly obvious that the Bible had no concern with astronomy. Since God had created both nature and the Scriptures, both necessarily had to be true: if an apparent conflict came about, the Bible should take second

place, not because it was any less valid, but because its meaning was often obscure or ambiguous, whereas mathematical
demonstrations were straightforward and final. Above all,
Galileo warned of how unwise it would be to fix a biblical
meaning in advance if it were unrelated to the spiritual cause
of salvation, for in time such an interpretation might be
shown to be false. From the theological standpoint there were
snags in this reasoning, and when they emerged, Galileo sat
down and wrote another, more fulsome "Letter to the Grand
Duchess Cristina" in which he tried to untangle them, noting,
in addition—probably at Castelli's prodding—that Saint
Augustine, in his long treatise *On the Literal Interpretation of
Genesis*, had admonished the faithful not to read Genesis too
narrowly, lest a mulish adherence to the letter of this difficult
text should cause a "scandal" to the Church. For the Inquisition, this layman's citation of Augustine must have seemed a
piece of effrontery. But the cardinals at the 1633 trial did not
need, as lawyers say, to "reach" that damning fact, since the
"Letter to Benedetto Castelli" answered to their purpose, and
they left the letter to Cristina alone, probably out of respect for
the Tuscan ducal family.

We do not know what Galileo or anybody really believed at
this period, since religious belief was prescribed by an autocracy and heresy was an actionable offense. If one had misgiv-
ings, one kept them to oneself, so it would be naive to take
religious ruminations penned in the papal realm or its client
territories at face value. The Inquisition's own records confirm
that many people harbored reservations and heretical beliefs:
before the Counter-Reformation, they had been much more
candid about them. Giorgio Vasari, author of the *Lives* of the
Italian artists, of 1550, tells us that Perugino, painter of innu-

merable sweetly pious altarpieces, "was a person of very little religion, and no one could ever make him believe in the immortality of the soul"; Perugino can hardly have been alone. The biblical text had also lost credence. If it was still overwhelmingly felt to be in some sense the word of God, the Reformation had led to numerous books being expunged from the Protestant biblical canon, and the Vulgate, the Roman Catholic Bible in Latin, had also been called into serious question. This text had been assembled from a collection of manuscripts and largely translated into Latin by Saint Jerome in AD 390–405. Yet Hebrew joined Latin and Greek as a widely studied language in sixteenth-century Italy, and by the end of the Council of Trent, which lasted from 1545 to 1563, Jerome's Vulgate was thought to be inaccurate and too often obscure: though declared canonical in 1546, an emended edition was called for. It took three pontifical commissions to produce a corrected text in 1588 and yet another in 1592, and the two editions differed at almost five thousand points, leading to heated criticism and some derision. Meanwhile, the whole idea of a sacrosanct text had fared rather poorly in European humanist circles, where the basic concepts of modern philology had struck root. That Moses, the supposed scribe of the Pentateuch, could certainly not have written some of it—he dies toward the end of Deuteronomy— had already been noted in the Babylonian Talmud, a fact further discussed by Abn Ezra, a twelfth-century rabbinical commentator much consulted by Christian exegetes, and this and similar problems had been quietly reviewed. Within a few years of Galileo's trial, a Catholic, Isaac La Peyrère, two Protestants, Thomas Hobbes and Samuel Fisher, and, later, a Jew, Benedict Spinoza, were to deny that humanity has an accurate copy of the true Bible, or that it was written by any but a number of authors.

So the Counter-Reformation was fighting a losing battle against a freer, and often less literal, reading of the Scriptures. Yet if there was one thing that had concerned the Council of Trent, it was the possibility that laymen would decide for themselves what passages in the Bible could be interpreted other than literally. In fact, the issue of the earth traveling about the sun had little if any bearing on the Catholic faith. But the notion that persons without theological training could decide for themselves to read this or that biblical passage in a nonliteral sense constituted a mortal danger for Catholicism in the early seventeenth century. The Protestant churches had split from Rome in denial of the literal meaning of the Sacrament of the Eucharist, reaffirmed in Session VIII of the Council of Trent, in 1551, as "truly, actually, and substantially the body and blood" of Christ. It places no great demand upon our sympathetic imagination to understand that the Church, in an age before the ideal of free expression had attained currency anywhere, was not about to allow laymen to decide whether phrases like "This is my body" (Mathew 26:26) were to be taken literally. A magnificent civilization had been rent asunder over just these issues. Consequently, though what Galileo said about Scripture was perfectly reasonable (indeed, a wise commentary on 2 Corinthians 3:6, "For the letter killeth, but the spirit giveth life"), it was also most impractical, reflecting a devout but overconfident sense of mission.

If the Inquisition's charges against Galileo were that he had championed Copernicanism conclusively rather than counterfactually, as a hypothetical argument, and that he had interpreted Scripture according to his own lights, then he was basically guilty. The serious mitigating factors—Bellarmine's certificate and the Church's ample opportunity to review the *Dialogue* before its printing—were not taken to heart. The entire trial focused on the issue of insubordination, with only three short paragraphs being devoted to the substance of Galileo's science; and these paragraphs, which occur in the report from the consultor Zaccaria Pasqualigo and concern the causes of the tides, merely show that Pasqualigo did not understand what he had read. Certain Galilean scholars have asserted that the cognitive issues raised by the conflict between Galileo's Copernicanism and the Church's views about astronomy and the Bible had already been taken up in the period between 1610 and 1616, and especially during the Inquisition's investigation of Galileo at the end of that period. With the best will in the world, I cannot agree: there are no records to support such a view. No Church document from 1615 to 1616 carefully examines either the scientific plausibility of heliocentrism or the complex issue of its compatibility with the Roman faith. The "Consultants' Report on Copernicanism" of February 24, 1616, in large measure a response to Galileo, states only that the heliocentric proposition is "foolish and absurd in philosophy" and "explicitly contradicts in many places the sense of Holy Scripture." Bellarmine's already mentioned letter to Father Foscarini of April 12, 1615, also intended for Galileo, represents the Vatican's broadest discussion of this topic, yet basically it dismisses the question out of hand. When, at the outset, Bellarmine says, "It appears to me that Your Reverence and Sig. Galileo have acted prudently in being satisfied with speaking in terms of assumptions," denoting with this term a merely suppositional Copernicanism, he is not complimenting Foscarini and Galileo on their intellectual tact. No: he is issuing a veiled, Jesuitical warning. Already the dynamic of intimidation has begun.

Did Galileo and the Church ever square off intellectually? The answer must be negative. Though initially acclaimed by the Collegio Romano, Galileo ran afoul of the Jesuits around 1624 with the "Reply to Ingoli" and *The Assayer*, which was to some degree an attack on Father Orazio Grassi. Both Ingoli and Grassi were churchmen who disputed various aspects of Copernicanism, Ingoli seriously and consistently and Grassi confusedly and tangentially. Neither represented the Vatican, and neither accused Galileo of heresy. Moreover, it is likely that Maffeo Barberini, as newly elected pope, supported Galileo against the Jesuits at the time; one recalls that *The Assayer* was reprinted with a dedication to Urban VIII. (Neither the "Reply" nor *The Assayer* was cited at the trial.) In fact, at various periods Galileo had devoted supporters inside the Church—not to mention Maffeo Barberini himself, one thinks of Sarpi, Castelli, Dini, and Ciampoli—and he had many ruthless antagonists in the secular academic community. It is only toward 1633, rather suddenly and sadly, that science and religion as social forces began identifiably to collide. What often goes unremembered is that the Church was then a temporal power with a police force (the papal *sbirri*) and a court system at its command. Thus the confrontation with Galileo, when it came, was not a genuine discussion but an act of coercion. The coercion was mild by modern standards— those, let us say, of the Moscow show trials of 1938—but it was coercion nevertheless. If there is any lesson to be derived from the Galileo affair, it is probably that the dialogue between science and religion, which is valuable on many, especially ethical, points, should not be contaminated by the use or threatened use of state power. The obedience to a hierarchy, the reliance on formalistic or ritualistic structures, the reference to sanctified interpretations of scriptural texts whose very spirituality derives from their fluidity and continual expansion of meaning—these aspects of the authoritarian posture destroy any hope of a positive exchange. Religion is often demagogically enlisted to rally a threatened social order, and such gambits, too, frustrate the intellectual probity of the debate.

There is another dimension to this trial that is seldom if ever mentioned, though it may be the most critical one. Historically the Italians have had little use for ideologies, which have tended merely to mask their mundane needs and wishes. In Galileo's period, as now, people's ambitions were tied to family allegiances, to the desire to further the fortunes and powers of one's *casa*, one's dynastic house. Little else mattered. The great political writing of the preceding decades— Machiavelli, Castiglione, Guicciardini—had assumed a clear-eyed cynicism on the part of the more capable princes and bishops, dismissing as hypocrisy all noble sentiment. One suspects, therefore, that a tissue of underhanded pressure, linked not to ecclesiastical but to family interests, underlay this important trial. Indeed the dominant presence of the Barberini, Medici, and Borgia clans suggests a mind-boggling maze of *clientelismo*, and who knows whether the famous plea bargain was not the consequence of somebody's "having the goods" on somebody else—for a few blessed weeks at least. Calling to mind this world of hushed antechambers and darkened sacristies, of signifying nods and whispered Latin words, we should be on our guard against the tendency to indulge in high-minded philosophical interpretations of the skimpy trial records that have survived from that spring of 1633. In the way of documents, however, they are, along with a few letters, all we have (unless more of the Barberini correspondence should come to light). If the usual Roman machinations took place, we don't know who may have been importuning or cajoling whom.

We have seen that between 1615 and 1632, the Vatican's position toward Galileo was exquisitely ambivalent, tending now toward favor, now toward hostility. This lay in the nature of autocracy, which exercises power most effectively if its subjects do not know exactly where they stand—their very uncertainty enforces awe and timidity. But what if a subject comes along whose ideas are both monumentally brilliant and potentially transgressive? For the prince to accept such ideas would diminish him by comparison, while their suppression would make him look stupid and cruel. This was the dilemma the pope faced in 1632–33.

Historians have two chief theories as to why Maculano's extrajudicial settlement collapsed, leading to a long and punitive trial. One was that a Jesuit cabal was launched against Galileo. The other was that a vindictive Maffeo Barberini intervened and axed the plea deal. Both may be true at once, and much of interest has been written outlining complex possibilities. One thing, however, cannot be denied: nothing could be done without the pope's consent.

In order to understand Maffeo, a cultivated and intelligent if vain and temperamental man, it is important to note that he had several different *kinds* of reasons to be angry at Galileo. First, he had personal reasons: this fellow Tuscan whom he had encouraged, whom he had addressed as "son" and "brother," and whom he admired rather as he admired

Bernini, had betrayed him in the *Dialogue* by assigning his theology to a weak-minded debater with the unfortunate monicker of Simplicio; the pope may also have thought that Galileo had enlisted Ciampoli to hoodwink him into sanctioning the book. Second, Maffeo had doctrinal reasons, relating to his pontifical role: Galileo had espoused heliocentrism and trespassed into biblical territory, which were the crimes for which the Inquisition had indicted him. Third, Maffeo may have had political reasons, resulting from his decisions as a temporal ruler. This is not the place to explain Pope Urban VIII's claims to theocratic absolutism after 1627; his disastrous meddling in the Mantuan succession; his support for a powerful Protestant leader, Gustavus Adolphus of Sweden; his befuddled attempts to arbitrate among the great Catholic powers; or his provocation of the Spanish Hapsburgs, who, it may be remembered, ruled the Kingdom of Naples, just south of the Papal States. In a secret consistory of March 1632, Cardinal Gaspare Borgia, the Spanish ambassador, had attacked the pope so vehemently that guards had had to be summoned: Maffeo was an inveterate francophile, and Rome was rife with rumors that the Spanish wanted him deposed. Ciampoli had been implicated in the disturbance, and the pope, as we have seen, always made much of Ciampoli's connection with Galileo (though certainly Galileo did not). It is hard not to surmise that, all things considered, Maffeo felt obliged to set in motion an imposing institutional machine that would demonstrate his control over the Galileo affair. A quick fix did not serve his publicity purposes.

But Galileo's ordeal and ultimate condemnation were the reverse of everything Maffeo's papacy, with its yearning to fuse Renaissance humanism and post-Tridentine piety, had stood for. It was a tragedy for him as well as for Galileo. Long ago he had written to the philosopher Tommaso Campanella about the 1616 decree: "It was never our intention; had it been up to us, that decree would not have been passed." But once again, as in the days of Pope Paul V and Cardinal Bellarmine, the Vatican had refused to see, refused to look up and perceive the world as it is, the heavens as they are made. It was a kind of blindness, as Kepler had so insightfully put it, though in the long run this blindness had little or nothing to do with optics, or astronomical representation, or any objection to Galileo's comfortable reliance on vision, on sense data. It had to do with the sheer difficulty of looking fixedly at what seems to be distressing or upsetting. Maffeo may have had lingering regrets about this. Years later, when Benedetto Castelli implored the pope, through his brother Antonio, to accept Galileo's assurance that he had never intended to hurt his feelings, Maffeo answered—rather sorrowfully, I think—"We believe him, we believe him."*

THE SENTENCE OF the Holy Congregation, handed down on June 22, convicted Galileo of holding and teaching Copernicanism, of interpreting Scripture according to his own meaning, and of deceitfully gaining permission to publish the *Dialogue*. Bellarmine's certificate was discounted. Galileo's crime was that of being "vehemently suspected of heresy," a religious offense situated between full-fledged heresy and

---

* It is in the nature of the Italian intellectual temper frequently to detect hidden, conspiratorial causes behind social events. In 1987 Pietro Redondi published an eminently readable and informative book titled *Galileo eretico* (*Galileo Heretic*), in which, among other things, he claimed that the real, previously undiscovered reason for the judicial investigation of Galileo in 1632–33 was not Copernicanism at all but the suspicion of heresy for atom-

temerity, or rashness. He was to be imprisoned indefinitely, though this part of the sentence was very soon commuted. He spent the rest of his life under house arrest at Arcetri, in the province of Florence.

His abjuration of Copernicanism, at the convent of Santa Maria sopra Minerva, paved the way for his penitence and—in the Vatican's view—his benign treatment. "I abjure, curse, and detest the above-mentioned errors and heresies," he said on his knees. Cardinal Francesco Barberini was not present. Also absent were Cardinals Borgia and Zacchia.

---

ism. Apparently Galileo's atomistic speculations in *The Assayer*, according to his Jesuit enemies, denied the miraculous nature of the Eucharist, the transformation of bread and wine into the body and blood of Christ during Communion. In fact, Redondi had discovered a document, the so-called *G3*, probably written by a Jesuit foe of Galileo's in 1632, which seemed to refer this matter to the Inquisition's attention. (More recently, Mariano Artigas has discovered an analogous, previously unknown report to the Holy Office known as the *EE291*.) Redondi's argument had the glamour of proposing that there was a head-on collision between science and the Church over the nature of matter itself, and, as in a mystery novel, that the clash had been disguised by a cover-up mounted by the Barberini clan, guilty of having admired *The Assayer* and backed Galileo for years. Few scholars supported Redondi's thesis for technical reasons pertaining to the nature of the documents in question, but one need not reach such grounds to reject the notion that Galileo could ever have been tried for the heresy of denying transubstantiation. First, a heresy charge entails the allegation of intentionality, and nobody could possibly have shown that Galileo had revived the ancient theory of atomism with the express intention of seducing his readers into heretical opinion. Second, what Galileo said about atoms, though it arguably ran counter to the Church's doctrinal, pseudo-Aristotelian explanation of the transformation of bread and wine into flesh and blood, had no demonstrable bearing on the *miraculous* nature of that transformation. It was, at worst, "erroneous."

Invidia

There is a moment in Galileo's *Dialogue* in which one of the participants says something that must, at the time, have sounded incredible, perhaps even profoundly disturbing. Reading this passage, you feel that you're standing at one of the turning points in the story of human thought. In this part of Galileo's fictional discussion, Salviati, the champion of the heliocentric theory, is talking about gravity and the apparent fact that bodies fall straight downward along a plumb line. "We observe the earth to be spherical," he says, "and therefore we are certain that it has a center, toward which we see that all its parts move. We are compelled to speak in this way, since their motions are all perpendicular to the surface of the earth, and we understand that as they move toward the center of the earth, they move toward their whole, their universal mother. Now let us have the grace to abandon the argument that their natural instinct is to go not toward the center of the earth, but toward the center of the universe; for we do not know where that may be, or whether it exists at all.

Even if it exists, it is but an imaginary point; a nothing, without any quality."

In rapid succession, then, Salviati has asserted that the center of the earth is not the center of the universe; that we don't know where that center is, or indeed if it exists; and that even if it does exist, it is just a point of no particular interest. This arresting assertion, which places the earth in the middle of nowhere, contradicts not only two thousand years of educated assumptions concerning its position, but also the teachings of the Catholic Church's most authoritative theologian, Saint Thomas Aquinas, and the vision of Italy's greatest poet, Dante Alighieri, whose *Inferno* (like much Christian folklore) situates Hell at the center of the earth, at the farthest distance from Heaven.

Salviati's remark is worth examining. Following on the heels of the expression "their universal mother," used with reference to the earth, it sounds at first like a metaphor. Generally he is a hard-headed customer, given to offering geometrical proofs for his assertions, but he proposes none for this one—it is as if he wants to shock his contemporaries out of their cosmological complacency, to float an image of a universe radically different from the one they conveniently imagine. Yet a moment's reflection is enough to convince us that this is not a metaphor. It has the magic of a metaphor, but it is a statement of fact.

Galileo, who was not only a scientist but also a trained musician, a prose stylist, and an occasional poet, always recognized the distinction between metaphors and factual assertions. As we have seen, one of his objections to Aristotle was leveled against the notion of the incorruptible firmament, an array of perfect heavenly bodies circling about the earth, as opposed to our corruptible, sublunary world. For Galileo, this lingo of corruptibility did not belong in "mathematical philosophy"—what we call science: it smacked of ethics or mythology. Despite his admiration for Ovid, Dante, and Ariosto, Galileo warned on several occasions against the contagion of science by thought processes proper to poetry or metaphysical speculation. Yet when he wrote about science, as opposed to doing it, and especially when he defended his ideas against potential attacks from theologians, he used many phrases with metaphorical undertones, others that might well be construed as metaphors, and still others that simply were metaphors.

Galileo had a keen comic sense. Elsewhere in the *Dialogue* Salviati talks about a man walking around the globe, noting that after a while his head will have traveled farther than his feet. For a moment we wonder whether we're reading a work by Edward Lear, and not a seventeenth-century scientific genius, until we realize that the statement is literally true. And Galileo can range farther still. In a famous letter to a friend, the former archbishop Piero Dini, of March 23, 1615 (that is, about two years after beginning to study the sunspots), Galileo wrote, "It seems to me that there is to be found in nature a most spirited, tenuous, and fast-moving substance which is diffused throughout the universe, which penetrates everything without resistance, and warms up, vivifies, and fecundates all living creatures. It seems to me that the senses themselves show that the main recipient of this spirit is the body of the sun, from which light is radiated throughout the universe, accompanied by that caloric spirit which penetrates all vegetable bodies, making them alive and fruitful." This brief meditation, with its undertone of mystical sun-worship and its suggestion that the universe is organically constituted, was not intended for public consumption; it was, perhaps, a flight of fancy. But Galileo's most famous trope, the idea of "the book of nature," was unquestionably launched as a promotional tool in his campaign to secure the Church's acceptance for his brand of astronomy. "Philosophy," he wrote in *The Assayer*, of 1623, "is written in this grand book, the universe, which stands continually open to our gaze. But it cannot be understood unless one first learns to comprehend the language and recognize the letters in which it is composed. It is written in the language of mathematics, and its characters are triangles, circles, and other geometric figures, without which it is humanly impossible to understand a single word of it. Without these, one wanders about in a dark labyrinth." Here Galileo champions a mathematical science that can produce results that are quantifiable at least in the sense that Euclidean geometry can entertain quantities; but he also goes farther than that. He suggests, enlisting a metaphor, that nature is a book, analogous to Scripture and created by God, who has given us minds in order that we may decipher and understand it. Like the Vulgate, which is written with the twenty-four characters of the Latin alphabet, this book has its language, which is that of mathematics, and so far the trope seems to work; but as we think about it, we might wonder what Galileo is really talking about, for typically this brand of Galilean rhetoric wavers between the metaphorical and the simply factual. After all, the universe is not "composed" in any language—such a metaphor is technically inapt. What is composed in the language of mathematics is the ongoing project, the "book," if you will, of scientific research, and this "book"—the ideal library of all we will ever know about nature—Galileo has conflated with the image (rather hard to summon up mentally) of the universe as an infinite text. So the promotional tag "the book of nature" (already a time-honored phrase in 1623, and which he sometimes also refers to as "the book of philosophy") can refer to the universe, or to science, or to both at once.

The Italian humanists of the cinquecento and early seicento were much more at home with metaphorical thought than we are. As dab hands at Latin they all knew the works of Cicero, Longinus, and Quintilian, which exhaustively catalogue the forms and uses of the classic tropes. A lot of Mannerist and Baroque literature, especially of the minor sort, actually amounts to a maze of hypertrophied metaphors, rather like Vicino Orsini's garden of stone monsters at Bomarzo, near Viterbo, which we know means something though we are hard pressed to figure out what. And Galileo, of course, was part of this culture: his own contribution to poetic literature contains, among other oddments, two lectures on Dante's *Inferno*, which are mathematical rather than literary, and a slender sheaf of poems known as *Le Rime*.

Galileo's "lessons" on the geometrical layout of the *Inferno*, which he presented at two consecutive meetings of the Florentine Academy in late 1587 or early 1588, were a young man's geometrical divertimento. They have no scientific interest, but they remind us of the poetic content of his life. By poetic content I mean the scenery of a man's or woman's yearnings, which may be connected to a lost love, a house under construction, a city from which one has been exiled— whatever gives meaning to one's deepest reveries. Of course some people's lives have no poetic content, for not everyone is given to self-communion; but Galileo was much preoccupied with the story of his life, which, in his mind, was the romance of a great natural philosopher, the tale of his glorious discoveries and of the defeat of his envious, slanderous enemies. And, in a curious way, it must be said that one could viably build a Baroque romance around the biography of Galileo. That the Pisan mathematician as a very young man should have given talks on an imaginary world, and later gone on to describe the solar system and to defend his views against the contention that they were imaginary—there is a poetic logic in this, as if he himself were living a metaphor. In his Dante lectures Galileo explicated a treatise "On the Site, Form, and Measure of the Inferno and of the Giants and Lucifer" by Antonio Manetti, a Florentine architect of the late fifteenth century, and in the figures he drew for the audience he made repeated use of conic sections. His demonstrations have not survived, but they

Botticelli's Drawing of Dante's Inferno

Galileo analyzed Dante's *Inferno* as a series of conic sections in two lectures to the Florentine Academy in 1587 or 1588.

must perforce have resembled regularized, geometrical versions of Botticelli's famous illustration of the Inferno.

Galileo's poetry, which was of the "occasional" variety,
shows the influence of Francesco Berni, the gifted satirist and
wit from Pistoia who died, in his mid-thirties, in 1535. The
mathematician's most interesting effort, written sometime
between 1589 and 1592, concerns the compulsory wearing of
the academic gown, in which a simile, far-fetched but not
atypical of the period, is drawn between certain sexual practices and the Aristotelian manner of reasoning. Reading these
verses, one suddenly perceives Galileo as an amphibian,
equally at home on the terra firma of science and in the fluvial
rapids of the metaphorical. Mario Biagioli has discussed what
he calls Galileo's "remarkable skills in emblematics," by which
he means his ability to harness the symbolic aspects of his discoveries and ideas for their promotional value at the Medici
court. This was no minor aspect of his social role. During this
period there were no endowed chairs for great scientists, no
multimillion dollar research grants, no positions on the
boards of cutting-edge engineering firms, no protection for
one's scientific discoveries, no Nobel prizes. Beset by financial
problems even after his appointment as court philosopher in
1610, Galileo had to defend himself against a horde of petty
detractors, intellectual-property bandits, and serious astronomical and theological opponents. That he was continually
assailed by a flock of spiteful, pecking "pigeons"—a term he
derived from the name of one of his chief Aristotelian opponents, Ludovico delle Colombe (*colombo* means "pigeon" in
Italian)—was almost literally true. But he parlayed this truth
into something else, something more gripping, more emo-
tionally laden—the myth of himself, the myth of Galileo.

That the myth of Galileo was founded on truth is borne out by the many warnings in the letters of Galileo's close friends concerning the envy and enmity of his rivals, which it would be tedious to enumerate here. Of particular interest, though, is the correspondence that he conducted with the Tuscan painter Ludovico Cigoli, who, it will be remembered, was employed in Rome in 1610–12, together with Domenico Passignano, on the decoration of the cupola of the Pauline Chapel in Santa Maria Maggiore. Cigoli and Galileo had a high regard and much affection for each other, though the former, as a sort of elder brother, did not hesitate to criticize the latter's writing style when it grew too pompous. Cigoli's letters reveal that he was embittered by the envy of other painters over whom he had been preferred for the cupola commission: "all my pleasure," he confessed, "is accompanied by so much bitterness." Only two of Galileo's letters have survived, to twenty-nine of Cigoli's, but much of the exchange clearly concerns the backbiting that both men faced: the painter, as the wiser and calmer of the pair, makes use of his vicissitudes to steady Galileo's nerves against the disparagement he has to endure, and to encourage him to persevere in his research and in his will to publish. Cigoli in 1607 had already painted a strange *Deposition* showing both the sun and the moon, which, as Eileen Reeves has persuasively argued, had distinct pro-Copernican implications, even hinting through pictorial means that to oppose heliocentrism meant turning one's face away from the world as God had created it.

In October of 1610, we find Cigoli reporting to Galileo in disgust that Father Clavius, the elderly chief astronomer of the Jesuit Collegio Romano, has told a friend of his that he laughs at the Jovian moons, and that to convince anyone of their existence it would be necessary to make a spyglass that would first create them, hey presto, and then show them, and that Galileo was welcome to his opinion and he to his. Cigoli also warns Galileo about the possible negative outcome of publishing the *Starry Messenger* in Italian, so that anyone can read and attack it. "And it also irritates them and they make a great fuss over [somebody else] having invented the spyglass . . . and I tell all this to you dear sir so that you may gird yourself and that your enemies may not find you ill-armed for your defense." In November he tells Galileo that he is proudly showing off his letters and not to lose heart, because "every beginning creates a difficulty for those who are hardened and have grown sclerotic in an opinion. Yet in the end the truth shall have its way." By January of 1611 he can report that Father Clavius has, despite himself, confirmed the existence of the Jovian moons, but by summer he warns him of the intrigues of his Aristotelian opponents (probably Ludovico delle Colombe's followers), who "plant mines behind your back," and suggests that he "publicly dispose of their opinions." A little later he gets wind of Clavius's insistence on the moon's absolute smoothness. "I've thought and thought about it," he says, "and I can find no other fallback for his defense than this, that a mathematician, however grand he may be, if he happens to have no [ideas], is not only half a mathematician, but also a man without eyes."

Toward autumn he announces that Passignano has received a telescope from Venice and is beginning to look at the sunspots, which "seem to be wandering within the body of the sun." From this point forward, for as long as he remains in Rome, he will forward considerable information to Galileo about the sunspots, including a diagram of the rudimentary camera obscura that Passignano has contrived, along with a concave green lens, to observe the sun. Touchingly, he has elected himself Galileo's research assistant. There are, however, more warnings about Ludovico delle Colombe, about *invidia*, or envy, and about the *maledicenti*, or "slanderers." A certain Archbishop Marzimedici, he says, is trying to get a preacher to draft a broadside on parchment against Galileo with reference to the movement of the earth, but Cigoli will warn his friend if anything untoward actually happens. For much of 1612 he forwards sunspot observations, including careful drawings, noting on one occasion that he has recently made twenty-six separate viewings. He also tells Galileo with barely concealed amusement that Passignano, by nature solemn and opinionated, now seems to be claiming that he has observed stellar parallax. In the autumn of 1612, however, his disgusted tone returns: he speaks of the Aristotelians seeking not truth but the preservation of entrenched social position, and of having had wind of Galileo's controversy with Scheiner. He returns again to the theme of *invidia*, and the ever-conspiring mob of evildoers and charlatans; some defamers accuse him of secretly painting in oil, not fresco; and he has noticed Galileo's hesitation to let Prince Cesi publish certain eloquent missives (which would become *Letters on the Sunspots*). On no account should Galileo let anyone talk him out of this. "Do it, do it, do it," he says, "and do not let your own cause down, any more than you have done in the past. Write the truth, without overdoing it and without currying adulation or yielding the field to fortune's whims; and do not slow your course for them whether they be pigeons or geese— laugh them off, Signor Galileo."

But Cigoli was sometimes dejected himself, and then he liked to say (with an exasperated puff, one supposes), "Cigoli

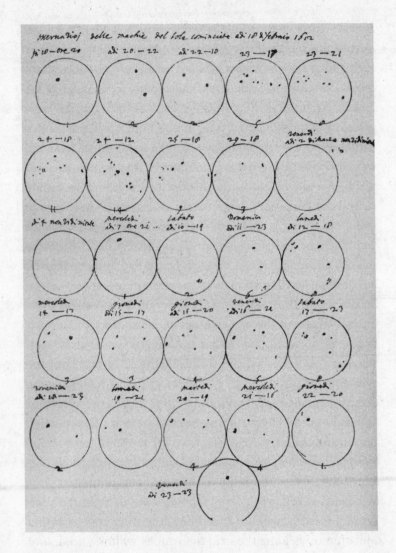

Cigoli's Sunspot Drawings

A page of sunspot drawings sent by Cigoli to Galileo early in 1612.

blows away with wind"—*Cigoli per il vento va via*. It was a wry pun that played his name off one of the most famous lines in Italian poetry,

> *E cigola per vento che va via*
> And hisses in the air that rushes away

which comes from the story of Pier della Vigna, in the *Inferno* (XIII, 42). This passage, which most educated Italians knew (and still know) by heart, tells the story of how Dante the protagonist, walking with Virgil through the Seventh Circle, comes upon the terrible Wood of the Suicides, and there, breaking a twig off a bush, hears the cry "Why do you tear me?" come hissing out of it, like air escaping from a burning branch. Dante's tale was based on sad reality. When alive, at the court of Frederick II of Naples, Pier della Vigna had been the victim of calumny, itself the product of the "whore" Invidia; envy had made him "unjust against my own just self," and he had taken his life in despair. Later, in *Purgatory* XIII, Dante describes the penance of the envious, who must endure having their eyes sewn tight with *fil di ferro*, "threads of iron"; and this may partly explain Cigoli's remark to Galileo about Father Clavius being "a man without eyes." In the Italian speech of this period, the term *invidia* often carries Dantesque connotations. The invidious refuse to see, or, when they do see, it is only to covet what is not theirs.

There is in the Uffizi a pair of drawings by Cigoli that Miles Chappell identified in 1975 as studies for an allegory of Invidia intended for Galileo, with reference presumably to what the

Cigoli's "Invidia"

A drawing by Cigoli showing Virtue or Truth menaced by Envy (Invidia). Cigoli saw both Galileo and himself as the targets of relentless social envy, hence this work, created with the mathematician in mind.

scientist was enduring at hands of his enemies and slanderers. The drawings show Virtue, in the form of a maiden of ideal aspect, overcoming Invidia, a hideous hag with Medusa-like hair who cowers beneath a boulder. In the drawings Virtue is half-metamorphosed into a tree or bush, which led Chappell to connect her, tentatively, with the nymph Daphne, who was changed into a laurel tree to evade violation by Apollo. But Daphne had no connection to virtue, nor Apollo to envy, and it is far more likely that Cigoli was thinking of the hissing sound of his own name and of poor Pier della Vigna. Most probably, then, these drawings emblematically depict Truth in the guise of a Pier della Vigna–like figure triumphing over the wicked persecution of Invidia. As images, they are the perfect embodiment of the dark, conspiratorial side of the Galileo myth.

Envy is the feeling of spite that others have what you do not have. In the catalogue of the sins it has several variations, chiefly jealousy, which is the fear that another will take or has already taken what is yours. Another variation, one might say, is pretense, or the claim that things are as you see them, and the refusal or incapacity to see them in any other, usually less flattering, way. Envy, jealousy, and pretense are all inspired by self-love. When Galileo and his supporters talk about Invidia, they are to some degree talking about the pretense of those who will not look at the world as God has made it, but only at the world as they, for their own vain purposes, would prefer it to be. In subtly warning Christians to avoid such contortions, Galileo was surely not thinking altruistically of the welfare of the Church: mostly he was aiming to protect his own scientific pursuits. But his argument was very sound advice.

NOTES

Prologue: The Summons

page

22 *The Assayer*: See Galileo Galilei, *Discoveries and Opinions of Galileo* (ed. and trans. Stillman Drake), Doubleday Anchor Books, Garden City, N.Y., 1957, pp. 256–258.

23 "failed to diminish": Ibid.

24 Erwin Panofsky and Eileen Reeves: See Erwin Panofsky, *Galileo as a Critic of the Arts*, M. Nihoff, The Hague, 1954; and Eileen Reeves, *Painting the Heavens: Art and Science in the Age of Galileo*, Princeton University Press, Princeton, N.J., 1997.

25 "the Tuscan artist": Noted by Reeves, *Painting the Heavens*, p. 13.

25 "encoded" or "indexed" in a subject's memory: I have culled these two examples from p. 486 of Richard Boyd's "Metaphor and Theory Change: What Is a 'Metaphor' a Metaphor for?" in Andrew Ortony, ed., *Metaphor and Thought*, Cambridge University Press, Cambridge, U.K., 1979, pp. 480–532. See also Ricardo Nirenberg, "Metaphor: The Color of Being," in Louis Armand, ed., *Contemporary Poetics*, Northwestern University Press, Evanston, Ill., 2007, pp. 153–174.

Galileo Galilei and Maffeo Barberini

28 Viviani's biography: Vincenzo Viviani's "Racconto istorico" is in vol. 19, pp. 597–632, of Galileo Galilei, *Le opere di Galileo Galilei*

(ed. Antonio Favaro), Barbera, Florence, 1929–39. Hereafter abbreviated *OGG*.

29 Aristotelian dynamics: For a discussion of the lack of an idea of force in Aristotle, see Paul Tannery, "Galileo and the Principles of Dynamics," in Ernan McMullin, ed., *Galileo Man of Science*, Basic Books, New York, 1968, pp. 163–177.

30 entomological illustration: For an example, see Stillman Drake, *Galileo at Work: His Scientific Biography*, University of Chicago Press, Chicago, 1978, p. 290.

30 "theory of the concave spherical mirror": Galileo's hypothetical ray diagram is particularly well illustrated in Edward R. Tufte, *Beautiful Evidence*, Graphic Press, Cheshire, Conn., 2006, pp. 80–81.

30 regarded Ariosto: All of Galileo's "Scritti letterari" are in *OGG*, vol. 9.

31 critical essay on Tasso's *Jerusalem Delivered*: This is the "Considerazioni al Tasso," in *OGG*, vol. 9, pp. 63–148.

31 anamorphosis: See the "Considerazioni," ibid., pp. 129–130.

31 taught perspective: Perspective, a drafting technique derived from geometrical optics, was part of the mathematics curriculum at many Italian universities in the late sixteenth century. There is to my knowledge no proof that Galileo taught it, but he referred to it in his letters to Ludovico Cigoli and he would have mastered it with ease.

31 elaborate treatises on the subject [of perspective]: Among others, *Comendarius di F. Commandinus*, Venice, 1558; Lorenzo Sirigatti, *Pratico della prospettiva*, 1596; Giordano Nunonario and Guidubaldo Del Monte, *Perspectivae Libri Sex*, Pesaro, 1600.

31 It has been noted: Panofsky, *Galileo as a Critic of the Arts,* and Reeves, *Painting the Heavens.*

33 As William R. Shea has pointed out: See William R. Shea, *Galileo's Intellectual Revolution*, Neale Watson Academic, New York, 1972, pp. 39–40.

33 Aristotle's cosmos: As elaborated in the *De Caelo*. See *On the Heavens* [Greek and English] (trans. W. K. C. Guthrie), Heinemann, London, 1939, esp. Book 2, pp. 130–255.

36 the mathematical concept of integration: See Galileo Galilei, *Dialogue Concerning the Two Chief World Systems—Ptolemaic &*

Copernican (trans. Stillman Drake), University of California Press, Berkeley, 1967, pp. 228–229. For the comparison of infinite sets, see Galileo Galilei, *Dialogue on the Two New Sciences* (1638) (trans. Henry Crew and Alfonso de Salvio), Macmillan, New York, 1914, p. 30ff.

40   Maffeo told Paul V: For this incident, I have noted: Herrera, "Memorie intorno la vita d'Urbano VIII," Barb. 4901, Biblioteca Apostolica Vaticana, pp. 48–50, which is cited in Ludwig [Freiherr] von Pastor, *The History of the Popes* (trans. Ernest Graf), Kegan, Paul, French, Trubner & Co., London, 1937, vol. 26, p. 387, n. 4; Herrera is cited again in Paolo Portoghesi, *Roma barocca: storia di una civiltà architettonica*, C. Bestetti, Roma, 1966, p. 54.

42   Maffeo's poetry: See *Maphaei S.R.E., Card. Barberini nunc Urbani PP VIII poemata*, ex typographia R. Cam. Apost., Romae, 1631. See also Lucia Franciosi, "Immagini e poesia alla corte di Urbano VIII," in Marcello Fagiolo, ed., *Gian Lorenzo Bernini e le arti visive*, Istituto della Enciclopedia italiana, Roma, 1987, pp. 85–90. Maffeo Barberini's poetry consists of paraphrases of the Psalms, sacred odes, meditations on the fugitive nature of life, and occasional and congratulatory verses. I confess that I haven't read much of it, though I have studied Franciosi's analysis. Of the third category, here is an example, slightly edited by me ("Poemata," p. 113):

> *Serio desiderium fugacis*
> *Vitae fascinat! Ut trahit voluptas!*
> *Ut cor abripit aura blanda plausus,*
> *Implicat laqueis opum cupido,*
> *Fallit Ambitio, tenetq[ue] luxus!*
> *Stulti quid sequimur [c]aduca? Fulgens*
> *Caeli Regiu nos vocat; sed armis*
> *Obniti Pietatis est necesse*
> *Contra nequitiae dolos.*

45   Maffeo was extraordinarily well disposed: For a detailed chronicle of this early affection, including the examples offered here, see Antonio Favaro, "Gli oppositori di Galileo: VI, Maffeo Barberini," *Atti del Reale Istituto Veneto di Scienze, Lettere ed Arti*, vol. 80, 1920–21, pp. 1–16, and vol. 81, pp. 17–46.

The Telescope; or, Seeing

53   Kepler, writing to Galileo on March 28, 1611: Letter 611 from M.
      Caspar et al., eds., *Johannes Kepler Gesammelte Werke*, C. H.
      Beckische Verlag, Munich, 1937, vol. 16, p. 372.

53   Stillman Drake has claimed: Stillman Drake, "Galileo's Steps to Full
      Copernicanism and Back," *Studies in the History and Philosophy of
      Science*, vol. 18, no. 1, 1987, pp. 93–105.

54   letter of 1597 to Jacopo Mazzoni: For a discussion, see Stillman
      Drake, *Galileo at Work: His Scientific Biography*, University of
      Chicago Press, Chicago, 1978, p. 40.

54   *ex suppositione*: For a full examination of the many forms of Aris-
      totelian suppositional reasoning (most not used by Galileo), see
      William A. Wallace, "Aristotle and Galileo: The Use of Hypothesis
      (*Suppositio*) in Scientific Reasoning," *Studies in Aristotle*, vol. 9, 1981,
      pp. 47–77.

54   Galileo wrote Kepler: On August 4, see *OGG*, vol. 10, pp. 67–68.

55   three lectures at the University of Padua: Only tiny fragments of these
      lectures have survived. See Drake, *Galileo at Work*, pp. 104–106.

56   Ronchi suggested a reason: Vasco Ronchi elaborated this thesis in his
      Nature of Light: An Historical Survey (trans. V. Barocas), William
      Heinemann, London, 1970.

57   vigorously disputed in 1972 by David C. Lindberg: David C. Lind-
      berg and Nicholas H. Steneck, "The Sense of Vision and the Origin
      of Modern Science," in Allen G. Debus, ed., *Science, Medicine, and
      Society in the Renaissance: Essays to Honor Walter Pagel*, Heinemann,
      London, 1972, vol. 1, pp. 29–45.

58   a focal length of 12 to 20 inches: I am citing Albert Van Helden's fig-
      ures. See Albert Van Helden, "The Invention of the Telescope," *Trans-
      actions of the American Philosophical Society*, vol. 67, no. 4, 1977, p. 11.

58   the first primitive spyglasses: See Engel Sluiter, "The Telescope
      before Galileo," *Journal for the History of Astronomy*, vol. 28, 1997,
      pp. 223–234, and Colin A. Ronan, G. L'E. Turner, et al., "Was There
      an Elizabethan Telescope?" *Bulletin of the Scientific Instrument Soci-
      ety*, vol. 37, 1993, pp. 2–10.

58   The first telescope: See Van Helden, "Invention of the Telescope," p. 42.

60   "a report reached my ears": Galilei, *Discoveries and Opinions*, pp. 28–29.

61   "News came": From *OGG*, vol. 6, pp. 258–259. Translated by Stillman Drake and C. D. O'Malley, *The Controversy on the Comets of 1618: Galileo Galilei, Horatio Grassi, Mario Guiducci, Johann Kepler*, University of Philadelphia Press, Philadelphia, 1960, pp. 212–213, quoted in Van Helden, "Invention of the Telescope."

61   "My reasoning": From *OGG*, vol. 4, pp. 258–259. Translated by Albert Van Helden, "Galileo and the Telescope," in Paolo Galluzzo, ed., *Novità celesti e crisi del sapere*, Giunti, Florence, 1984, p. 152.

62   Joining a bitter race: Recounted in Galileo to Beneditto Landucci, August 29, 1609, in *OGG*, vol. 10, p. 253.

62   Galileo wrote to the doge: See *OGG*, vol. 10, p. 250.

66   This description of the Galilean telescope is largely derived from the extraordinarily clear presentation by Tom Pope and Jim Mosher, "Galilean Telescope Homepage," available at www.pacifier.com/~ tpope/Galilean_Optics_Page.htm. Last accessed July 27, 2008.

67   a letter from Arcetri to Fortunio Liceti: From *OGG*, vol. 18, p. 233, quoted in Vasco Ronchi, *Il Cannocchiale di Galileo e la scienza del Seicento*, Edizioni Scientifiche Einaudi, Torino, 1958, p. 139.

68   Galileo discussed magnification very warily: See "The Starry Messenger," in Galilei, *Discoveries and Opinions*, p. 30. In the passage beginning "Now in order to determine . . . the magnifying power of an instrument . . .," he is forthcoming enough to offer the reader the following information: "Let [the user] draw two circles or two squares on paper, one of which is four hundred times larger than the other . . . [that is, has twenty times its width]. He will then observe from afar both sheets fixed to the same wall, the smaller one with one eye applied to the glass and the larger one with the other, naked eye. This can easily be done with both eyes open at the same time. Both figures will then appear of the same size if the instrument multiplies objects according to the desired proportion [x20]." This is all very well and good, but it reveals nothing of how to make a telescope! It is so obvious that it seems addressed to readers who have never even possessed a pair of reading glasses.

68   His competitors: See Mario Biagoli, *Galileo's Instruments of Credit: Telescopes, Images, Secrecy*, University of Chicago Press, Chicago, 2006, p. 93ff.

69   "And I shall describe it": Letter from Giovanni Battista Della Porta to Federico Cesi, August 28, 1609, from *OGG*, vol. 10. Translated by Van Helden, "Invention of the Telescope," p. 44.

71   Ippolito Francini: See Albert Van Helden, *Catalogue of Early Telescopes*, Istituto e Museo di Storia delle Scienze/Giunti, Florence, 1999, p. 30.

71   The composition of Galileo's lenses: The information offered here is based on my interview with Dr. Giorgio Strano of the Institute and Museum of the History of Science, Florence.

74   The very large dusky patches: From "The Starry Messenger," in Galilei, *Discoveries and Opinions*, p. 31.

75   "The surface of the Moon": Ibid., p. 31.

75   to Antonio de' Medici: From *OGG*, vol. 10, pp. 273–278.

76   "Now on Earth": From "The Starry Messenger," in Galilei, *Discoveries and Opinions*, p. 33. For the geometrical proof of the minimal height of a lunar mountain cited here, see ibid., pp. 31–42.

80   C. W. Adams found: C. W. Adams, "A Note on Galileo's Determination of the Height of Lunar Mountains," *Isis*, vol. 17, 1932, pp. 427–429.

81   his long letter: From *OGG*, vol. 10, pp. 273–278.

82   he wrote his friend Belisario Vinta: From *OGG*, vol. 10, p. 280.

83   Galileo soon concluded that they were moons: Galilei, *Discoveries and Opinions*, p. 57.

83   he manufactured many hundreds of telescopes: See Biagoli, *Galileo's Instruments of Credit*, pp. 90–94.

89   At last, in October of 1610: For dating, see letter of December 30 to Benedetto Castelli, 1610, in *OGG*, vol. 10, pp. 502–504.

89   If his earlier observations: For the *Letters on Sunspots*, see Galilei, *Discoveries and Opinions*, pp. 59–85.

90   "I confess to your Excellency": Ibid., p. 113.

91   Carrington: To picture what Richard Carrington discovered, remember that the earth rotates from west to east. Then imagine a cloud that while also rotating west to east has a proper motion, a motion of

its own, much slower than the rotation of the earth. With respect to longitude, its proper motion is east to west.

92   "I seem to have observed": Ibid., p. 130.

93   axis of the sun's tilt: Ibid., p. 125.

93   "to verify the rest of the [Copernican] system": Ibid., p. 144.

93   "perhaps this planet also": Ibid., p. 144.

93   Cesare Cremonini: See Biagioli, *Galileo's Instruments of Credit*, p. 113.

94   Giulio Libri: See Drake, *Galileo at Work*, p. 162.

94   Martin Horky wrote a letter to Kepler: From *OGG*, vol. 10, pp. 142–143.

94   the classical literary etymology of the Italian word for envy, *invidia*: With respect to envy, Charles S. Singleton writes (in *Dante Alighieri: The Divine Comedy: Inferno: 2. Commentary*, Bolligen Series LXXX, Princeton University Press, Princeton, NJ, 1970, p. 213: "The sin of envy is thought of as movement of the eyes, first of all. Thus, in *Purg.* XIII, souls are purged of that sin by having their eyelids sewed shut. Pietro di Dante, commenting on *Purg.* XIII, says: 'Invidia facit, quod non videatur, quod expedit videre; et ideo dicitur *invidia*, quasi *non visio*.' ('Envy causes that which should be seen not to be seen. And therefore is called *invidia*, almost as if to say, nonvision.') [Thus also] the *Magnae derivationes* of Uguccione da Pisa: 'Invideo tibi, idest non video tibi, idest non fero videre te bene agentem.' ('I envy you—that is to say, I do not see you; that is, I cannot bear to see you doing so well.')" This etymology goes back to the Latin grammarian Priscian (ca. AD 500) and beyond him to Cicero.

94   "I must write of a harsh objection": Quoted in Ronchi, *Cannocchiale di Galileo*, p. 139. For *cronicatori* I read "star chroniclers," that is, those who note the exact times of the rising and setting of the stars.

95   the problem of confirmation: Biagioli, *Galileo's Instruments of Credit*, pp. 27–44, 132.

96   a primitive camera obscura: This whole issue is best described in ibid., p. 192, n. 141.

97   public sessions: See ibid., pp. 86–90.

97   Galileo's wash drawings: See *OGG*, vol. 10, p. 274ff.

99   which Galileo apparently had a large hand in: See Horst Bredekamp, *Galilei der Kunstler: Der Mond. Die Sonne. Die Hand*, Akademie Verlag, Berlin, 2007, p. 189ff, under the heading "Galilei als Stecher?" Bredekamp, having examined the hatching employed in the shading of the lunar craters and compared it with Galileo's drawings, writes, "One is allowed to suspect that Galileo, pressed by time as the book neared production, enjoyed a phase as a graphic artist" (my translation).

99   Gugliemo Righini: For a discussion, see M. L. Righini Bonelli and William R. Shea, eds., *Reason, Experiment, and Mysticism in the Scientific Revolution*, Macmillan, New York, 1975, pp. 59–76.

103   Father Christopher Clavius: See Reeves, *Painting the Heavens*, p. 151.

103   The other tradition: Of the many works discussing the doctrine of the immaculate conception, see especially Jaroslav Pelikan, *Mary through the Centuries: Her Place in the History of Culture*, Yale University Press, New Haven, Conn., 1996, pp. 177–200.

105   Gallanzone Gallanzoni: A *cavaliere*, or knight, from Rimini, he was at this time Cardinal Joyeuse's secretary. See Galileo's letter to Gallanzone of July 16, 1611, in *OGG*, vol. 11, p. 143, and the commentary in Reeves, *Painting the Heavens*, pp. 17–18, 216–220, who reads it as an "implicit criticism of the emerging doctrine of the Immaculate Conception, the basis of many of the Marian associations with the moon."

106   "Having crossed that fiery sphere": This as yet unpublished translation of Astolfo's moon voyage is by David Slavitt. See Ludovico Ariosto, *Orlando furioso* [Canto Trentesimoquarto], Successori Le Monnier, Firenze, 1888, pp. 254–256.

108   "They have, as it were": Johannes Kepler, *Conversations with the Sidereal Messenger* (trans. and ed. Edward Rosen), Johnson Reprint Corporation, New York, 1965, p. 28.

109   painted dome: See John Shearman, "The Chigi Chapel in S. Maria del Popolo," *Journal of the Warburg and Courtauld Institute*, vol. 24, 1961, p. 138ff, and Philippe Morel, "Morfologia delle cupole dipinte da Correggio a Lanfranco," *Bolletino d'arte*, ser. 6, vol. 69, no. 23, 1984, pp. 1–34.

111   *invidia* implied: See Reeves, *Painting the Heavens*, p. 17.

111   Cigoli, then living in Rome, wrote Galileo: From OGG, vol. 10, pp. 290–291.

112   Passignano: See again Biagioli, *Galileo's Instruments of Credit*, p. 192, n. 141; also Reeves, *Painting the Heavens*, p. 5.

117   Samuel Edgerton: Samuel Y. Edgerton, *The Renaissance Rediscovery of Linear Perspective*, Basic Books, New York, 1975, p. 162. Giotto's acquaintance with Alhazen's optics, long suspected, has received further confirmation by Giuliano Pisani in the January 2008 issue of *Bolletino del museo civico di Padova*, widely reviewed in the Italian press. I have not, however, been able to obtain a copy.

119   decree condemning Copernicanism: Maurice A. Finocchiaro, ed. and trans., *The Galileo Affair: A Documentary History*, University of California Press, Berkeley, 1989, pp. 146–150.

119   Bellarmine warned him: Ibid., pp. 147–148.

121   Bellarmine stated: Translated by Richard J. Blackwell in *Galileo, Bellarmine, and the Bible: Including a Translation of Foscarini's Letter on the Motion of the Earth*, University of Notre Dame Press, Notre Dame, Ind., 1991, p. 266.

121   "Considerations on the Copernican Opinion": This can be found in Finocchiaro, *Galileo Affair*, pp. 70–86.

122   These are circular sophisms: See Guido Morpurgo-Tagliabue, *I Processi di Galileo e l'epistemologia*, Armando, Rome, 1981, pp. 51–59.

123   Francesco Ingoli: See *OGG*, vol. 5, pp. 403–412. Ingoli was made a consultor to the Holy Office and recognized as a quasi-official anti-Copernican (and by extension anti-Galilean) critic only *after* writing this notably weak polemic. See Annibale Fantoli, *Galileo, for Copernicanism and for the Church* (trans. George V. Coyne), Vatican Observatory Publications, Vatican City, 1996, p. 255, n. 50.

123   Galileo politely refuted: See "Galileo's Reply to Ingoli (1624)" in Finocchiaro, *Galileo Affair*, pp. 154–197.

124   Stillman Drake's phrase: See Galilei, *Discoveries and Opinions*, p. 264, n. 14.

128   as William R. Shea has written: Shea, *Galileo's Intellectual Revolution*, p. 163.

129   "Now when we see this beautiful order": Galilei, *Dialogue*, p. 367.

130   Shea has termed: Shea, *Galileo's Intellectual Revolution*, p. 163.

130   a different graphic representation: See Galilei, *Dialogue*, pp. 342–345.

130   Mount a crossbow: Ibid., p. 168ff.

131   the church of San Petronio: Ibid., p. 463. I have seen this discussed in detail only by J. L. Heilbron in his fascinating *The Sun in the Church: Cathedrals as Solar Observatories*, Harvard University Press, Cambridge, 1999, pp. 176–180. I thank my friend Norman Derby, professor of physics, for his help in understanding Professor Heilbron's enlightening but somewhat confusing diagram.

132   as Shea has observed: Shea, *Galileo's Intellectual Revolution*, p. 181.

The Trial; or Not Seeing

136   In 1575, as the Counter-Reformation: All this and more information concerning Tuscany's increasing pliability with respect to the Papal States can be found in Furio Diaz, *Il Granducato di Toscana: I Medici*, UTET Libreria, Turin, 1976, Part III, "La Toscana nell'età della Controriforma," pp. 274–278, 287–288, 321–323, and 323–326.

138   Ciampoli's verse: See Giovanni Ciampoli, *Poesie sacre*, Carlo Zenero, Bologna, 1648. This collection consists of verses on the utility of sacred poetry; meditations based on the Psalms; and songs of praise, notably for the Santa Casa di Loreto. There is also a "Cantico delle Benedittioni" for the coronation of Pope Urban VIII. For an interpretation, see Franciosi, "Immagini e poesia alla corte di Urbano VIII."

139   "calumny of": From *OGG*, vol. 14, pp. 383–385.

139   "I began to think, as you so rightly say": Ibid., pp. 388–399.

140   "In such affairs of the Holy Office": Ibid.

142   "*Basta, basta!*": Letter of September 18, from ibid., pp. 391–393.

142   "speak cautiously": Ibid.

142   "my great esteem for you": From *OGG*, vol. 14, pp. 118–119.

144   The missive: Ibid., pp. 406–410.

146   he wrote Galileo a shrewd note: Letter of October 23, from ibid., pp. 418–419.

147   "I tried to awaken in him": Letter of November 13, from ibid., pp. 427–428.

148   harvesting hundreds of thousands of lives: Lodovico Antonio Mura-
       tori, in his classic *Annali d'Italia: Dal principio dell'era volgare sino
       all'anno MDCCXLIX*, Classici Italiani Contrada del Cappuccio,
       Milan, 1820, vol. 15, p. 117, gives the figure 560,000, including
       500,000 for the terra firma of the Veneto and 60,000 for the other
       northern and central Italian regions. He does not offer
       documentation.

150   Benedetto Castelli: Letter of November 20, from *OGG*, vol. 14, pp.
       430–431.

152   to his old friend Elia Diodati: From *OGG*, vol. 10, January 15, pp.
       23–26.

153   "Though the affairs of this tribunal": February 16, 1633, from *OGG*,
       vol. 15, p. 41.

153   He wrote Cioli on February 19: From *OGG*, vol. 10, pp. 43–45.

154   "in the guise of a visitor": February 19, 1633, from *OGG*, vol. 15, pp.
       43–45.

154   "I think Serristori": Ibid.

155   He decided to write two cardinals: From *OGG*, vol. 15, pp. 46, 49.

155   "From what I gather": Niccolini to Cioli, February 27, from *OGG*,
       vol. 10, pp. 54–55.

157   "At the Palace": I have used James J. Langford's translation (from
       Galileo, Science, and the Church, Desclee, New York, 1966), quoted in
       Richard J. Blackwell, *Behind the Scenes at Galileo's Trial: Including the
       First English Translation of Melchior Inchofer's Tractatus syllepticus*,
       University of Notre Dame Press, Notre Dame, Ind., 2006, p. 5.

157   Galileo wrote to a friend: To Geri Bocchineri, February 25, from
       OGG, vol. 10, p. 50.

158   *amorevolezza*: Niccolini to Cioli, February 27, from ibid., p. 55: "*non
       mancano chi dubiti che difficilmente [Galileo] habbia a scansar d'esser
       ritenuto al S. Offizio, bensí si proceda seco sin adesso con molta
       amorevolezza e placidità.*"

159   And indeed Francesco conceded: Niccolini to Cioli, February 27,
       from ibid., pp. 55–56.

159   "may God forgive": Niccolini to Cioli, March 13, from ibid., pp.
       67–68.

160   As a temporal prince: On foreign ambassadors visiting Urban VIII, see

Leopold von Ranke, *History of the Popes: Their Church and State* (trans. E. Fowler), Colonial Press, New York, 1901, vol. 2, pp. 371–374.

161 Maffeo's response: Letter of March 13, from *OGG*, vol. 15, pp. 67–68.

161 On April 9, only three days before the trial: Niccolini to Cioli, April 9, from *OGG*, vol. 10, p. 84.

162 I am indebted to Professor Christopher F. Black for pointing out to me that Galileo was almost certainly not interrogated in Latin, as the record (not a perfect verbatim transcript) would suggest. See also pp. 63–68 of Black's authoritative *The Italian Inquisition*, Yale University Press, New Haven, Conn., 2009.

164 The Roman Inquisition has hardly been studied: The principal source used here on the legal practices of the Roman Inquisition is John Tedeschi, *The Prosecution of Heresy: Collected Studies on the Inquisition in Early Modern Italy*, Medieval & Renaissance Texts & Studies, Binghamton, N.Y., 1991. For early Tuscan resistance to the Inquisition, see p. 92; see also p. 126ff; for the influence of the Roman legist Ulpian, see p. 143.

167 the *corda*: In the concluding chapter of his magisterial two-volume work, *La Tortura giudiziaria nel diritto comune* (Giuffré, Rome, 1954), Piero Fiorelli notes (pp. 231ff) that Cicero, Quintilian, Saint Augustine, Ulpian, Boccaccio, and Montaigne had all inveighed against the the use of judiciary torture. By 1633, it was regarded by many theologians and philosophers as an indefensible practice, and two important works by Jesuit fathers had bitterly criticized it: *Universa teologia scholastica* (1627) of Adam Tanner, a Tyrolese, and *Cautio criminalis* (1631) of Friedrich von Spee, a German.

167 "Torture," went a canon-law maxim: Tedeschi, *Prosecution of Heresy*, p. 144.

168 canon-law guidelines: For these, see L. Garzend, "Si Galilée pouvait être juridiquement torturé," *Revue des questions historiques*, vol. 90, 1911–12, pp. 353–389, and vol. 91, 1911–12, pp. 36–67.

168 the trial transcript for June 16, 1633: Sergio M. Pagano, ed., with Antonio G. Lucciani, *I Documenti del processo di Galileo Galilei* [Contro Galileo Galilei], Pontificiae Academiae Scientiarum, Vatican City, 1984, p. 154. Hereafter abbreviated *DPGG*.

170 Galileo at his first deposition: Ibid., pp. 124–130.

171   "I was notified": My translation. Ibid., p. 127.

172   "We, Robert Cardinal Bellarmine": Translated by Blackwell, *Behind the Scenes at Galileo's Trial*, p. 9.

172   Galileo described: Galileo's first deposition is found in *DPGG*, pp. 124–130.

176   On April 16, Niccolini wrote Cioli: From *OGG*, vol. 10, pp. 94–95.

176   He wrote Cioli a jubilant note on April 23: From ibid., p. 103.

176   The day before, in a letter discovered only in 1999: Translated by Blackwell, *Behind the Scenes at Galileo's Trial*, p. 14.

177   In another letter: Ibid., p. 14.

178   "was of the same [heliocentric] opinion": This letter of October 2, 1632, is cited in Fantoli, *Galileo, for Copernicanism and for the Church*, p. 407. Here Fantoli convincingly cites it as evidence of the "disparity of views that existed among the Church authorities themselves."

180   Two days after Maculano's fateful meeting with Galileo, the accused gave his second deposition: *DPGG*, pp. 130–132.

183   "I haven't told him everything yet": Letter of May 22, from *OGG*, vol. 10, p. 132.

183   On May 10, Maculano: *DPGG*, pp. 135–137.

184   As Maurice Finocchiaro has emphasized: Maurice A. Finocchiaro, *Retrying Galileo*, University of California Press, Berkeley, 2005, p. 11.

185   At the end of the final session: From *DPGG*, pp. 154–155.

185   "I am here in your hands": *del resto, son qua nelle loro mani, faccino [sic] quello gli piace [sic]*.

186   what Richard J. Blackwell has called: See Blackwell, *Galileo, Bellarmine, and the Bible*, p. 37. See chapter 3 for a full analysis.

189   Perugino: See Giorgio Vasari, *The Lives of the Artists* (trans. Julia Conway Bondanella and Peter Bondanella), Oxford University Press, New York, 1991, p. 266.

189   concepts of modern philology: For the beginnings of modern biblical criticism, see Richard H. Popkin, "Spinoza and Bible Scholarship," in Don Garrett, ed., *The Cambridge Companion to Spinoza*, Cambridge University Press, Cambridge, U.K., 1986, pp. 383–407.

191   The "Consultants' Report on Copernicanism": A translation can be found in Finocchiaro, *Galileo Affair*, p. 146.

191   "It appears to me": Translated by Blackwell in *Galileo, Bellarmine, and the Bible*, p. 266.

196   "It was never our intention": Quoted in Favaro, "Gli oppositori di Galileo," p. 18.

196   "We believe him": Ibid., p. 39.

196   "It is in the nature": See Ernan McMullin, *The Church and Galileo*, University of Notre Dame Press, Notre Dame, Ind., 2005, pp. 191–233.

Epilogue: Invidia

198   a moment in Galileo's *Dialogue*: Galilei, *Dialogue*, p. 36–37.

200   a famous letter: Translated by Blackwell in *Galileo, Bellarmine, and the Bible*, p. 206.

201   "Philosophy . . . is written in this grand book": Galilei, *Discoveries and Opinions*, pp. 237–238.

202   "lessons" on the geometrical layout of the *Inferno*: From *OGG*, vol. 9, pp. 31–57.

204   The mathematician's most interesting effort: From ibid., pp. 213–223.

204   Mario Biagioli has discussed: See Mario Biagioli, *Galileo Courtier: The Practice of Science in the Culture of Absolutism*, Chicago University Press, Chicago, 1993, pp. 107–159.

205   Cigoli: This account of the Galileo-Cigoli correspondence is based on the letters as they appear in *OGG*, vol. 10, pp. 241, 243, 290, 441, 456, 475, 478; and vol. 11, pp. 36, 132, 167, 175, 208, 212, 228, 241, 268, 286, 290, 318, 347, 361, 369, 386, 410, 418, 424, 475, 484, 501.

205   Eileen Reeves: Reeves, *Painting the Heavens*, pp. 138–183.

209   a pair of drawings by Cigoli: See Miles Chappell, "Cigoli, Galileo, and *Invidia*," *Art Bulletin*, vol. 57, no. 1, March 1975, pp. 91–98.

Selected Bibliography

Original Sources

Galilei, Galileo. *Le opere di Galileo Galilei* (ed. Antonio Favaro), 20 vols. Barbera, Florence, 1929–39. Abbreviated *OGG*.

Pagano, Sergio M., ed., with Antonio G. Lucciani, *I Documenti del processo di Galileo Galilei* [Contro Galileo Galilei]. Pontificiae Academiae Scientiarum, Vatican City, 1984. Abbreviated *DPGG*.

Books

Ariosto, Ludovico. *Orlando furioso* (trans. David Slavitt). Harvard University Press, Cambridge, in press.

Aristotle. *On the Heavens* (trans. W. K. C. Guthrie). Heinemann, London, 1939.

Banfi, Antonio. *Vita di Galileo Galilei*. Cultura, Milan, 1930.

Barberini, Maffeo. *Maphaei SRE, Card. Barberini nunc Urbani PP VIII poemata*. R. Cam. Apost., Rome, 1631.

Biagioli, Mario. *Galileo Courtier: The Practice of Science in the Culture of Absolutism*. Chicago University Press, Chicago, 1993.

———. *Galileo's Instruments of Credit: Telescopes, Images, Secrecy*. University of Chicago Press, Chicago, 2006.

Black, Christopher F. *The Italian Inquisition*. Yale University Press, New Haven, Conn., 2009.

Blackwell, Richard J. *Galileo, Bellarmine, and the Bible: Including a Translation of Foscarini's Letter on the Motion of the Earth*. University of Notre Dame Press, Notre Dame, Ind., 1991.

———. *Behind the Scenes at Galileo's Trial: Including the First English Translation of Melchior Inchofer's Tractatus syllepticus*. University of Notre Dame Press, Notre Dame, Ind., 2006.

Bredekamp, Horst. *Galileo der Kunstler: Der Mond. Die Sonne. Die Hand*. Akademie Verlag, Berlin, 2007.

Chappell, Miles L., ed. *Disegni di Ludovico Cigoli*. Exhibition catalogue. Olschki, Florence, 1992.

Ciampoli, Giovanni. *Poesie sacre*. C. Zenero, Bologna, 1648.

Deo Feo, Vittorio, and Vittorio Martinelli, eds. *Andrea Pozzo*. Electa, Milan, 1996.

Diaz, Furio. *Il Granducato di Toscana: I Medici*. UTET Libreria, Turin, 1976.

Drake, Stillman. *Galileo at Work: His Scientific Biography*. University of Chicago Press, Chicago, 1978.

Edgerton, Samuel Y. *The Renaissance Rediscovery of Linear Perspective*. Basic Books, New York, 1975.

———. *The Heritage of Giotto's Geometry: Art and Science on the Eve of the Scientific Revolution*. Cornell University Press, Ithaca, N.Y., 1991.

Evans, Robin. *The Projective Cast: Architecture and Its Three Geometries*. MIT Press, Cambridge, 1995.

Fagiolo, Marcello, ed. *Gian Lorenzo Bernini e le arti visive*. Istituto della Enciclopedia italiana, Rome, 1987.

Fantoli, Annibale. *Galileo, for Copernicanism and for the Church*. Vatican Observatory Publications, Vatican City, 1996.

Finocchiaro, Maurice A., ed. and trans. *The Galileo Affair: A Documentary History*. University of California Press, Berkeley, 1989.

Finocchiaro, Maurice A. *Retrying Galileo*. University of California Press, Berkeley, 2005.

Fiorelli, Piero. *La Tortura giudiziaria nel diritto comune*, 2 vols. Giuffré, Rome, 1954.

Galilei, Galileo. *Discoveries and Opinions of Galileo* (trans. and ed. Stillman Drake). Doubleday Anchor Books, Garden City, N.Y., 1957.

Galilei, Galileo. *Dialogue Concerning the Two Chief World Systems— Ptolemaic & Copernican* (trans. Stillman Drake). University of California Press, Berkeley, 1967.

———. *Sidereus nuncius, or, The Sidereal Messenger*. Chicago University Press, Chicago, 1989.

———. *Le Rime* (ed. Antonio Marzo). Salerno, Rome, 2001.

Gingerich, Owen. *The Eye of Heaven: Ptolemy, Copernicus, Kepler*. American Institute of Physics, New York, 1993.

Haskell, Francis. *Patrons and Painters: A Study in the Relations between Italian Art and Society in the Age of the Baroque*. Knopf, New York, 1963.

Heilbron, J. L. *The Sun in the Church: Cathedrals as Solar Observatories*. Harvard University Press, Cambridge, 1999.

Kemp, Thomas. *The Science of Art: Optical Themes in Western Art from Brunelleschi to Seurat*. Yale University Press, New Haven, Conn., 1990.

Kepler, Johannes. *Conversations with the Sidereal Messenger* (trans. and ed. Edward Rosen). Johnson Reprint Corporation, New York, 1965.

Kirwin, William Chandler. *Powers Matchless: The Pontificate of Urban VIII, the Baldachin, and Gian Lorenzo Bernini*. P. Lang, New York, 1997.

Lindberg, David C., and Robert S. Westman. *Reappraisals of the Scientific Revolution*. Cambridge University Press, Cambridge, U.K., 1990.

McMullin, Ernan, ed. *The Church and Galileo*. University of Notre Dame Press, Notre Dame, Ind., 2005.

Morpurgo-Tagliabue, Guido. *I Processi di Galileo e l'epistemologia*. Armando, Rome, 1981.

Muratori, Lodovico Antonio. *Annali d'Italia: Dal principio dell'era volgare sino all'anno MDCCXLIX*, 14 vols. Classici Italiani Contrada del Cappuccio, Milan, 1820.

Nussdorfer, Laurie. *Civic Politics in the Rome of Urban VIII*. Princeton University Press, Princeton, N.J., 1992.

Onori, Lorenza Mochi, Sebastian Schütze, and Francesco Solinas, eds. *I Barberini e la cultura europea del Seicento*. De Luca Editori d'Arte, Rome, 2007.

Panofsky, Erwin. *Galileo as a Critic of the Arts*. M. Nihoff, The Hague, 1954.

Pasquali, Giorgio. *Storia della tradizione e critica del testo*. F. Le Monnier, Florence, 1962.

Pastor, [Freiherr] Ludwig von. *The History of the Popes* (trans. Ernest Graf), Kegan Paul, French, Trubner & Co., London, 1937.

Pedrotti, Frank L., Leno M. Pedrotti, and Leno S. Pedrotti. *Introduction to Optics*. Pearson/Prentice Hall, Upper Saddle River, N.J., 2007.

Pérez-Gómez, Alberto, and Louise Pelletier. *Architectural Representation and the Perspective Hinge*. MIT Press, Cambridge, 1997.

Prosperi, Adriano. *L'Inquisizione romana: letture e ricerche*. Edizioni di storia e letteratura, Rome, 2003.

Ranke, Leopold von. *History of the Popes: Their Church and State* (trans. E. Fowler), 3 vols. Colonial Press, New York, 1901.

Redondi, Pietro. *Galileo Eretico*. Einaudi, Turin, 1983.

Reeves, Eileen. *Painting the Heavens: Art and Science in the Age of Galileo*. Princeton University Press, Princeton, N.J., 1997.

Ronchi, Vasco. *Il Cannocchiale di Galileo e la scienza del Seicento*, Edizioni Scientifiche Einaudi, Turin, 1958.

———. *The Nature of Light: An Historical Survey* (trans. V. Barocas). William Heinemann, London, 1970.

Shea, William R. *Galileo's Intellectual Revolution*. Neale Watson Academic, New York, 1972.

Shea, William R., and Mariano Artigas. *Galileo in Rome: The Rise and Fall of a Troublesome Genius*. Oxford University Press, New York, 2004.

Taton, R., and C. Wilson, eds. *Planetary Astronomy from the Renaissance to the Rise of Astrophysics: Part A. Tycho Brahe to Newton* [*The General History of Astronomy*]. Cambridge University Press, Cambridge, U.K., 1989.

Tedeschi, John. *The Prosecution of Heresy: Collected Studies on the Inquisition in Early Modern Italy*. Medieval & Renaissance Texts & Studies, Binghamton, N.Y., 1991.

Van Helden, Albert. *Catalogue of Early Telescopes*. Istituto e Museo di Storia della Scienza/Giunti, Florence, 1999.

Wlassics, Tibor. *Galilei critico letterario*. Longo Editore, Ravenna, 1974.

Articles and Chapters

Adams, C. W. "A Note on Galileo's Determination of the Height of Lunar Mountains." *Isis*, vol. 17, 1932, pp. 427–429.

Beretta, Francesco. "Le Procès de Galilée et les Archives du Saint-Office." *Revue des sciences philosophiques et théologiques*, vol. 83, 1999, pp. 441–490.

Cajori, Florian. "History of the Determination of the Heights of Mountains." *Isis*, vol. 12, 1929, pp. 482–512.

Campanella, Tommaso. "The Defense of Galileo" (trans. Grant McColley). *Smith College Studies in History*, vol. 22, nos. 3–4, April–July 1937.

Chappell, Miles. "Cigoli, Galileo, and *Invidia*." *Art Bulletin*, vol. 57, no. 1, March 1975, pp. 91–98.

"Cigoli, Ludovico [Ludovico Cardi detto il Cigoli]," in *Bollettino della Accademia degli Euteleti della città di San Miniato*, S. Miniato, Florence, n.d.

Drake, Stillman. "Galileo's Steps to Full Copernicanism and Back." *Studies in the History and Philosophy of Science*, vol. 18, no. 1, 1987, pp. 93–105.

Favaro, Antonio. "Gli oppositori di Galileo, VI; Maffeo Barberini." *Atti del Reale Istituto Veneto di Scienze, Lettere ed Arti*, vol. 80, 1920–21, Part 2, pp. 1–46.

Garzend, L. "Si Galilée pouvait être juridiquement torturé." *Revue des questions historiques*, vols. 90 and 91, 1911–12, pp. 353–389 and pp. 36–67.

Giacchi, Orio. "Considerazioni giuridiche sui due processi contro Galileo," in *Nel Terzo Centenario della morte di Galileo: saggi e conferenze*. Università Cattolica del Sacro Cuore, Vita e Pensiero, Milan, pp. 383–406.

Lindberg, David C., with Nicholas H. Steneck. "The Sense of Vision and the Origins of Modern Science," in Allen G. Debus, ed., *Science, Medicine and Society in the Renaissance: Essays in Honor of Walter Pagel*. Heineman, London, 1972, vol. 1, pp. 29–46.

Morel, Philippe. "Morfologia delle cupole dipinte da Correggio a Lanfranco." *Bolletino d'arte*, ser. 6., vol. 69, no. 23, 1984.

Popkin, Richard H. "Spinoza and Bible Scholarship," in Don Garrett, ed., *The Cambridge Companion to Spinoza*. Cambridge University Press, Cambridge, U.K., 1996, pp. 383–407.

Ronan, Colin A., G. I.'E Turner, et al. "Was There an Elizabethan Tele-scope?" *Bulletin of the Scientific Instrument Society*, vol. 37, 1993, pp. 2–10.

Sluiter, Engel. "The Telescope before Galileo." *Journal for the History of Astronomy*, vol. 28, 1997, pp. 223–234.

Van Helden, Albert. "The Invention of the Telescope." *Transactions of the American Philosophical Society*, vol. 67, no. 4, 1977, pp. 5–64.

———. "Galileo and the Telescope," in Paolo Galluzzo, ed., *Novità celesti e crisi del sapere*. Giunti, Florence, 1984, pp. 149–158.

———. "The Telescope and Authority from Galileo to Cassini." *Osiris*, 2nd ser., vol. 9, "Instruments," 1994, pp. 8–29.

Van Helden, Albert, and Mary Winkler. "Representing the Heavens: Galileo and Visual Astronomy." *Isis*, vol. 82, no. 2, June 1992, pp. 195–217.

Illustration Credits

Index

Page numbers in *italics* refer to illustrations.

Index

10. For a complete discussion of this topic, see the author's book *Fight Like a Man*, available from www.abbafather.com.

CHAPTER 10: THE FATHER AND THE MAN: OF FATHERS AND SONS

1. Bob Baum, "Family First: Ainge quits as Suns coach," *Santa Barbara News Press* (December 12, 1999), C1.

CHAPTER 11: THE FATHER AND THE MAN: OF FATHERS AND DAUGHTERS

1. This teaching is available in both audio and video from the author's Web site.
2. Jonetta Rose Barras, *Whatever Happened to Daddy's Little Girl?* (New York: Ballentine, 2000).
3. Wendy Shalit, *A Return to Modesty* (New York: Free Press, 1999).
4. Linda Warren, "Soul Sisters," *Charisma*, April 2003, 53.
5. Kathleen Hendrix, "$10-Millon Woman," *Los Angeles Times,* October 4, 1991, E1, 4.
6. Germaine Greer, *Daddy, We Hardly Knew You* (New York: Knopf, 1989).
7. *Women Respond to the Men's Movement*, edited by Kay Leigh Hagan (San Francisco: Pandora, 1992), vii.
8. Robert M. Herhold, *The Promise Beyond the Pain* (Nashville: Abingdon, 1979), 77.
9. Madeleine L'Engle, *A Wrinkle in Time* (New York: Dell, 1962), 152–200.

CHAPTER 12: TO KNOW THE FATHER

1. Chaim Potok, *The Chosen* (New York: Simon & Schuster, 1967).

CHAPTER 13: WHERE ARE ALL THE MEN?: WHY MEN DON'T GO TO CHURCH

1. Sharon Mielke, "Church analyst says trends need serious attention: Low number of men and big churches is major problem in UMC, district and conference leaders told," *United Methodist Reporter* (September 10, 1982), 1.
2. Leon J. Podles, *The Church Impotent: The Feminization of Christianity* (Dallas: Spence, 1999). This particular quote from Podles, "Missing Fathers of the Church," *Touchstone* magazine (January/February 2001), 26.
3. Ed Robb, "Is the Church Feminized? An Interview with Dr. Donald Joy," *Challenge to Evangelism Today* (July/August 1983, vol. 16), 1.

4. Dr. Lee Salk, *My Father, My Son* (New York: G. Putnam's Sons, 1982).

5. Anne S. White, *Trial by Fire* (Kirkwood, MO: Impact Books, 1975), 102.

6. *Alcoholics Anonymous Big Book* (New York: Alcoholics Anonymous World Services, Inc., 1976), 59.

CHAPTER 14: RATIONAL AND INDEPENDENT, FAITHLESS AND ALONE

1. David W. Smith, *The Friendless American Male* (Ventura, CA: Regal Books, 1983), 50.

2. Carol Gilligan, *In a Different Voice* (Cambridge, MA: Harvard University Press, 1982), 6–10.

3. Ibid.

4. Eugene Peterson, "Introduction to Galatians," *The Message* (Colorado Springs, CO: Navpress, 1993), 340.

EPILOGUE: THE MIRROR OF TRUTH

1. I heard this story some years ago at a conference, and have been unable to find its source.